AF573921

EARLY Scientific INSTRUMENTS

EARLY Scientific INSTRUMENTS

BY NIGEL HAWKES

INTRODUCTION BY D. J. BRYDEN
Curator, Whipple Museum of the History of Science,
University of Cambridge

ABBEVILLE PRESS
PUBLISHERS • NEW YORK

FRONT COVER
Equinoctial Sundial (18th century)
Commentary on page 23

TITLE PAGE
Rose Engine Lathe
Commentary on page 137

PICTURE CREDITS
Kenneth Ullyett Collection: p. 31. National Maritime Museum, Greenwich: pp. 27, 34, 37, 38, 41, 42, 48, 51, 52, 71. Royal Institution, London: pp. 93, 96–97, 99, 101, 123, 125, 126, 129, 143. Science Museum, London: all other pages. All photographs by Michael Holford.

Library of Congress Cataloging in Publication Data

Hawkes, Nigel.
Early scientific instruments.

Bibliography: p. 164
1. Scientific apparatus and instruments—History.
I. Title.
Q185.H35 681'.75'09 80-70462
ISBN 0-89659-192-1 AACR2

 Inquiries should be addressed to Abbeville Press, Inc., 505 Park Avenue, New York, N.Y. 10022. Printed and bound in Japan.

Design: Roy Winkler

CONTENTS

INTRODUCTION

During the early decades of the Renaissance, scholars slowly accepted that to practice science was to observe the natural world. To assist this examination of nature, various artifacts were developed to help the unaided senses. In time, careful and controlled observation, experiment, and measurement became the accepted approach to scientific knowledge.

The oldest of the sciences, astronomy, has roots which can be traced even beyond classical antiquity. By the beginning of the present millennium, astronomers had begun to use a few simple devices to aid them in following the motions of the sun, moon, and stars. By the Middle Ages the sophisticated astrolabe had become the symbol of man's growing comprehension of the complexities of the universe. In the late fifteenth century, new navigational instruments, their design based on astronomical theory, enabled intrepid explorers such as Christopher Columbus and Vasco da Gama to venture into uncharted seas with confidence.

In 1608 an Italian professor of mathematics, Galileo Galilei, turned a hitherto unexploited Dutch optical device to the skies. He revealed a vast multitude of new stars unseen by the naked eye. Four satellites circled the planet Jupiter; our moon, no longer a unique object in the heavens, was seen to have mountains and valleys. Even the perfect disc of the sun was blemished with spots. Conventional wisdom was shattered by this artifact, a small tube holding two lenses. Almost overnight, the cherished belief in a finite and earth-centered universe, bounded by God in his heaven, was demolished by Galileo and the newly invented telescope.

At the opposite end of the scale of creation, the compound microscope showed seventeenth-century man a new and totally unexpected world, the world of minutiae. Here was an instrument able to reveal nature's smallest secret places. The half century that followed Galileo's great discoveries with the telescope saw the invention not only of the microscope, but also of such devices as the logarithmic slide rule and mechanical calculator, the barometer, thermometer, air pump, and the pendulum clock. This was the period that historians now call the Scientific Revolution, when new instruments played vital roles in the development of a radically different understanding of natural phenomena. The spectacular advances of modern science and its cousin, technology, all date from this time. A major factor in these advances was the fruitful interplay between the studious scholar and the careful craftsman; the essential inspiration of the theoretician was matched by the equally vital technical skills of the artisan.

In 1471 the astronomer and mathematician Johannes Müller, called Regiomontanus, settled in Nuremberg, a town then at the center of the major European trade routes. At that time Nuremberg's guilds contained many craftsmen whose skill in fine metalwork was unrivaled. Regiomontanus chose to live in this city-state "on account of the availability of instruments, particularly the astronomical instruments on which the entire science of the heavens is based." With the encouragement of such men, Nuremberg and its neighbor Augsburg became famous for their scientific craftsmen, whose intricate astrolabes, fine quadrants, and sundials of amazing ingenuity provide incontrovertible evidence of their skill. By the late sixteenth century, however, the workshops of Nuremberg were beginning to be matched by others, responding to the needs of scientists in cities as far afield as Florence, Louvain, and Prague; and the disruptions of the Thirty Years' War (1618–1648) irrevocably damaged the pre-eminence of the German cities.

If Nuremberg and Augsburg set the standard for the sixteenth century, it was London that dominated the eighteenth. By this time the product range of the instrument-maker had widened considerably. Up to and including the sixteenth century, instruments were largely associated with the science of astronomy and its practical offshoot, gnomics—the telling of time from the position of the sun, moon, or stars. By 1700, precision instruments were used by a variety of "mathematical practitioners," men whose professional duties required the application of principles enuciated by mathematicians, geometricians, and astronomers. In addition, the new natural philosophy of the previous century had spawned a host of new instruments. The major innovations—telescope, microscope, barometer, therometer, and air pump—played vital roles in the development of a new approach to science. The number of different devices was large enough in 1700 for an enterprising author to compile the first dictionary devoted to scientific instruments. It comprised over one hundred entries, from almacanther staff to waywiser.

The leading instrument-maker in eighteenth-century London was Jesse Ramsden, who developed the circular dividing engine. This machine was capable of marking out mechanically the scales of measuring instruments, such as the theodolite and the sextant, with an accuracy and speed far in excess of that attainable by manual methods. Until his death in 1800, Ramsden led an intense and competitive search for the combination of quality and quantity in production. More typical of the London entrepreneurs of this century was Benjamin Martin, who began his career as a schoolmaster and itinerant lecturer on science. Martin made his reputation as a popular lecturer, explaining science to the gentry with the aid of an imaginative array of illustrative apparatus. Eventually he became the proprietor of a Fleet Street instrument-making business, and grew sufficiently renowned to be entrusted with the re-equipping of Harvard College after a fire destroyed all its scientific apparatus in 1764.

The eighteenth-century advance from instrument-making as a handicraft to a trade, where individual craft skills are harnessed to the overall requirements of a large workshop, was pioneered in London. German firms were responsible for the next step. They led the nineteenth-century development of precision mechanical and optical engineering as the trade became an industry. The major figure in this advance was Georg Reichenbach, a Bavarian infantry officer who in 1804 co-founded the Mathematische-Mechanisches Institut von Reichenbach, Utzschneider, und Liebherr. To complement the Mechanical Institute at Munich, Utzschneider, a wealthy lawyer, funded an associated glassworks and optical workshop, managed from 1809 to 1826 by Joseph von Fraunhofer. Fraunhofer's research was to earn him a permanent memorial as discoverer of the Fraunhofer Lines in solar spectra. As a technician, his skill in optical glassmaking and lens grinding put precision optics on a par with precision mechanics. Through the founding of a workshop in Jena by Carl Zeiss in 1847, German optical instruments became world leaders, and so they have remained.

During the eighteenth century, London exported both its wares and its talents. Catherine the Great staffed the newly established instrument workshops at St. Petersburg with craftsmen from London. The Royal Swedish Academy of Sciences sent journeymen to London to learn the finer arts of the trade. In this century London-trained artisans were to be found making instruments not only in the provincial capitals of Dublin and Edinburgh but further afield, in Amsterdam, Paris, Lisbon, and a number of towns in the New World.

Initially the North American colonists relied on imports of London-made instruments, but among the settlers were men trained in the trade and others who mastered the skills required to manufacture the basic but essential devices used in surveying and navigation. Because brass was in short supply, these practical men worked in the native woods: cherry, apple, oak, and maple. In the following century, Yankee ingenuity was stimulated by the immigration of highly skilled Germans, who helped establish the indigenous American instrument-making industry on a sound basis. They were particularly prominent in the field of optical instruments: Jacob Bausch and Henry Lomb, founders of the present-day internationally respected giant combine of Bausch & Lomb, were immigrants who left Germany in the turbulent 1840s. Joseph Zentmayer of Philadelphia and the Grunow brothers of New Haven were also German refugees of that generation.

The science of electricity has dominated the development of scientific instruments over the last 150 years. Widespread interest in electrical phenomena followed the discovery in 1745-46 by Ewald van Kleist of Cammin and (independently) Pieter von Musschenbroek of Leyden that a glass flask (the Leyden Jar) could be used to store the electric charge produced by an electrostatic generator. The next fifty years saw a plethora of apparatus and experiments, as the "electricians" tried to discover the unexpected properties of the "electric fluid." The great American statesman and diplomat Benjamin Franklin was a leading pioneer in the electrical investigations of the latter half of the eighteenth century. His "one-fluid theory" was used to explain electrostatic phenomena until the time of Faraday. Franklin also demonstrated, through a famous but somewhat dangerous experiment, that lightning was an electrical phenomenon; and in 1749 he proposed the introduction of lightning conductors, putting the idea into practice three years later.

In 1800 Alessandro Volta announced the discovery of energy in a new form: a constant flow of electricity from a chemical source, the voltaic pile, or battery. With Volta's battery, electricity moved from the erratic spark produced by the friction generator toward its present position of essential value to much of human endeavor. That magnetism could produce electricity was separately demonstrated in 1831 by Michael Faraday in London and Joseph Henry in Albany. Electromagnetic induction proved to be the key to a new electrical technology, which first flowered in the invention of the electric telegraph. It also provided a useful tool for physicists trying to understand the nature of matter.

Another simultaneous and independent discovery occurred in 1836: Father Nicholas Callan of Maynooth College in Ireland and Dr. Charles Page of Salem designed the first induction coil, which rapidly became the principal source of high-voltage electric current. Daniel Ruhmkorff, a German instrument-maker working in Paris, perfected the induction coil during the mid-1850s. At the same time, Johann Geissler, a specialist glass blower and mechanic at the University of Bonn, designed a new and very efficient vacuum pump. The "electric egg," a fascinating toy used by eighteenth-century "electricians," became the laboratory vacuum tube, with which physicists began to investigate the effects of high-voltage discharge through a very rarefied gas. In 1895 Konrad Roentgen discovered X-rays through the use of a vacuum tube, and a new diagnostic tool was added to the physician's armory. Less than two years later, J. J. Thompson, head of the Cavendish Laboratory at Cambridge, found that the rays emanating from the cathode of a vacuum tube were negatively charged "corpuscles." The electron, the key to modern physics, had been found.

In this book are illustrated important examples from the diverse range of artifacts with which man has tried to discover and explain the complexities of the physical world, and, through this comprehension, use nature for his own ends. These early instruments, in addition to providing a tangible record of the development of scientific knowledge, vividly demonstrate the technical ingenuity of former times. Present-day instrumentation is largely hidden behind a faceless computer console or a massive control panel, a mass of buttons, dials, and switches with flashing visual displays and the incessant clatter of printout. In startling contrast, these scientific relics reflect the confident calm of a lost age. Their design embodies the simple beauty of functional form.

—D. J. Bryden

EARLY
Scientific
INSTRUMENTS

David
Bouquet A Londres

This rare and unusual English table clock was made by David Bouquet of London, about whom little is known other than that he was admitted to the Clockmakers' Company in 1632 and died in 1665. It is a spring-driven striking clock with a verge escapement and a single hour hand, and would normally have been contained inside a case. This view shows the elaborate decoration and the maker's signature on the back of the clock.

Clocks like this were luxuries, expensive to buy and frivolous in the sense that they were poor timekeepers compared with long-case clocks of the same period. But they provided the convenience and freedom of mobility; they could be moved from room to room, and many were carried to the bedroom at night. Because they were portable they were more easily damaged than lantern clocks, which hung on the wall, or long-case clocks, which necessarily remained fixed where they were. They also broke easily, because the untempered metal used for the springs would often break at its anchorage. Few have survived.

Table clocks were certainly not utilitarian in design; even the back of this clock, which would only have been seen when the doors of the case were opened, is gloriously decorated and finished. The backplates of table clocks grew even more decorative in Georgian times, when they migrated from the table to the mantel, because there was invariably a mirror in the mantel behind the clock in which its splendid backplate could be reflected. In general, plain backplates persisted up to the 1680s, and the decorated backplate only began to become popular around 1675. Bouquet's clock is perhaps transitional; its backplate is ornamental, but not as elaborate as they subsequently became.

MERIDIES
EVRO✱NOTVS
LIBO✱NOTVS
APHRICVS✱LIBO
ZEPHYRVS
CIRCI✱VS
SEPTEN✱TRIO
AQVILO
VVLTVR✱NVS
SVBSOLANVS
CAPRICORNVS
DELPHIN
AQVILA
PEGASVS
ANDROMEDA
CAP.SERP.
CORONA
SCORPIVS
VENTER CETI
NARIS CETI
TAVRVS GEMINI CANCER LEO VIRGO
SPICA VIR
CANIS MAIOR
CANIS MINOR
HIDRA COR LEONIS
PATERA

The astrolabe was one of the most versatile instruments of the early astronomers. It is not primarily an instrument for observing the heavens, but an analog computing device which can be used for a large number of calculations. It can work out, for example, the time of rising or setting of any star, the length of the day or night, the time, and the time and duration of twilight, as well as the direction from which to expect any imminent celestial event.

The astrolabe is really no more than a two-dimensional representation of the stars in the vault of the heavens. There are two separate plane projections. The first, the *rete* or *spider*, is an openwork star map, with each little pointer representing the position of a star. The *rete*, which is normally the most attractive and distinctive feature of an astrolabe, can be rotated above the second plane projection, known as the *plate* or *tympan*. The *plate* represents in two dimensions a projection of the celestial sphere: engraved lines, arcs, and circles are the meridian, local horizon, equator, and tropics, with lines of azimuth and circles of altitude. This second projection varies with latitude; therefore, the body of an astrolabe will normally hold a number of plates, each engraved for a different latitude.

This astrolabe, made by George Hatman of Nuremberg in 1548, also has an *alidade*, the pivoted arm lying across the face of the instrument. This has small sighting holes in the vertical sections which are raised above the plane of the instrument. If the astrolabe is held vertically by the ring at the top, the *alidade* can be used as a simple sight through which the user looks at a star. The elevation of the star can be measured by using the fixed scale that runs around the outside of the instrument. From such an observation, the time of day, or which sign of the zodiac was in the ascendant, could be calculated.

XXIII
XXII
XXI
XX
XIX
XVIII
XVII
XVI
XV
XIIII
XII
VIII
VII

The classical historian Herodotus credited the Babylonians with the invention of the gnomon, a vertical rod the tip of whose shadow indicates the time. The science of gnomics was founded in the second century A.D., after Ptolemy had shown how the hour lines of a sundial could be constructed geometrically. With the flowering of mathematics in the Renaissance, sundials were made with considerable sophistication, in a huge variety of shapes and sizes.

This sundial was made in Florence in about 1560, and was kept in the Pitti Palace. It is made of wood, painted with distemper, and richly decorated with arabesques. It is a good example of the virtuosity of the period: five of its six faces tell the time according to different conventions.

The dial on the top of the cube shows Babylonian hours, in which the day consisted of twenty-four equal hours beginning at sunrise. On the north and south faces of the cube are two sundials which show the time according to old Italian hours, in which the day was divided into twenty-four hours beginning and ending at sunset. The east and west faces show the time in the ancient Jewish system, which divides the day into two periods, sunrise to sunset and sunset to sunrise, and then divides each of these periods into twelve hours. The length of these planetary or temporary hours varied depending on the time of year, and whether it was night or day.

To tell the time accurately, sundials must be designed specifically for the latitude of the location where they are to be used. This sundial is designed to work at a latitude of 46°, the latitude of Florence; unlike other sundials of the period, it has no provision for adjustment. It would therefore not be a very useful timepiece anywhere but Florence.

During the sixteenth and seventeenth centuries, a very popular form of sundial was the diptych dial, named after the Latin and late Greek word for a pair of folding writing tables. Small enough to be carried in the pocket, such dials were in effect the predecessors of the watch. Indeed, even after the watch was introduced, a sundial of some form or other was required to set it to the correct time. Between the two tablets of the diptych dial is a sloping wire or string, which is used to cast a shadow onto the dial, the large circle on the horizontal tablet. Since a sundial of this sort will only tell the right time if it's pointed in the right direction, the instrument is provided with a magnetic compass for orientation.

This diptych dial, made of ivory, is dated 1648. Stamped in the compass bowl is a fleur-de-lis, the master sign of Leonhart Miller of Nuremberg. The motto on the lower table, *Soli Deo Gloria* ("to God alone be the Glory"), is frequently found on Nuremberg dials of this period. Diptych dials were often richly engraved and decorated, with every spare corner used to convey useful information.

The string that connects the two tablets can be adjusted to five different positions, representing five latitudes between 42° and 54°, and there are five corresponding hour circles drawn on the horizontal tablet. A table on the vertical tablet gives the latitude of twenty important towns where the owner of the dial might find himself. In addition to the main sundial, there are two smaller dials, which tell the time in Italian hours (*Die Welsch Uhr*) and Babylonian hours (*Die Grose Uhr*), and a dial on the lid which indicates the length of the day (*Quantitas Diei*) or the position of the sun in the zodiac.

DANTZIG
LVBECK
PRAGA
ANTVERP
PARIS
NORNBERG
MAILAND
VENETIA
ROMA
NEAPOLIS
STETIN
LITAV
LVNDRA
COLONIA
VIENNA
AVGSPVRG
TRENT
LION
PORTVGAL
CONSTANTON
SEPT
SOLI DEO GLORIA
DIE GROSE VHR

IV
V
VI
VII
VIII
30 20 10
80 70 60

In 1574 a practical Elizabethan, William Browne, serving as Port-Reeve and Gunner on the bulwarks at Gravesend, noted that the equinoctial dial "be not used amongst our Mariners heere in England." He enthusicially praised this little instrument, "whiche is very profitable to knowe the houre of the day by in all latitudes through the whole worlde."

Sundials of this type have their pointers aimed directly at the North Star, which lies on the axis of rotation of the earth. When a rod is aimed in this direction, the sun rotates around it in a regular movement during the day; and whatever the date or the latitude, the shadow of the rod will point in the same direction at any given time of the day.

In use the sundial is oriented north-south by the compass in the base, and set horizontal by adjusting the four screw feet until the plumb bob hangs vertical. Then the hour circle is set to an angle equal to 90° minus the local latitude. When this has been done, the hour circle is parallel to the plane of the equator and the pointer is parallel to the axis of rotation of the earth. The sun appears to rotate around the pointer in a regular motion, producing a linear scale which is easy to calibrate on the hour circle. Each hour takes up 15°, or one twenty-fourth of the full circle.

Universal equinoctial dials were made throughout Europe from the seventeenth to the mid-nineteenth centuries. In Augsburg there was a distinct group of craftsmen, "Kompass-Macher," who specialized in making them. This example is German, possibly from a Berlin workshop, and was made in the second half of the eighteenth century. The stylized oak-leaf border engraved on the base plate is a characteristic decoration on instruments of this period.

LB
BOTH
BEARS

Nocturnals, or nocturlabes as they were sometimes called, are devices for telling the time at night. Their operation is based on the fact that the stars, while remaining fixed relative to one another, appear to rotate about the North Star. It is, of course, the earth which is rotating, and the North Star remains fixed because it lies along the Earth's axis of rotation. All the other stars rotate, and their position at any moment indicates the time.

A nocturnal consists of several pieces of metal, or in this case wood, pivoted together at the center so that they can rotate relative to one another. At the axis of rotation there is a hole, and in use the nocturnal is held upright by the handle until the North Star can be sighted through the hole. The long arm of the device is then turned until it lies along the line made by the two brightest stars in the constellation of the Great Bear (also known as the Big Dipper). These two stars are often used as "pointers" because they are easily seen and they lie along a line which passes close to the North Star. As an alternative, the bright star of the Little Bear (Little Dipper) can also be used.

This nocturnal is made to the design favored by London nautical instrument-makers in the early eighteenth century. It can be used with either the Great or the Little Bear, as noted by the legend "Both Bears" on the handle. If the Little Bear is used, the small pointer marked LB is turned until it lies against the date of the day on which the observation is being made; this automatically corrects the time measured from sidereal time to mean solar time (the solar day is an average four minutes longer). The North Star is sighted through the hole and the long arm turned until the bright star in the Little Bear lies on it. The time is then read off from the scale on the circumference on the central circle—just as if the long arm were acting as the hand of a clock.

If the Great Bear is chosen as a reference, the procedure is the same, except that the small pointer marked GB (partly hidden here) is set against the date.

Before the invention of clocks in the Middle Ages, a variety of ingenious methods of marking the passing of the hours had been devised. Charles V of France kept in his chapel a slow-burning candle marked with the hours; later, oil lamps with calibrated glass reservoirs were used as a rough measure of time. One of the very earliest devices was the clepsydra, a water vessel with a small hole in the bottom through which water flowed at a steady rate.

From the clepsydra it was a short step to the hourglass, in which sand takes the place of water. The hourglass was probably invented in the Mediterranean region in the twelfth or thirteenth century A.D. This one is a sixteenth-century English glass, about five inches high, with a stand made of wood, crudely carved and painted red. The twine around the waist of the glass gives the best clue to its age; few hourglasses were signed or dated, and the only way to identify them is by their method of manufacture.

Until the end of the eighteenth century, the hourglass was made of two separate vials, each drawn into a narrow neck at the point where they were to be joined. A diaphragm of thin sheet brass, drilled with a small hole, was placed between them, and then the whole construction was joined with wax or putty and bound with canvas and finally cord, thread, or fine wire. Unfortunately, the joint tended to work loose, and the penetration of moisture often caused stoppages in these early glasses.

Hourglasses were made to run for various periods of time: kitchen glasses commonly for three minutes, like today's egg-timers, and naval or watch glasses for half an hour or an hour. This glass runs for a quarter hour. Because the hole in the metal diaphragm tended to be worn by the flow of the sand and get progressively larger, sandglasses of this period were only approximate timekeepers. In 1703, it is said, a French fleet of five warships under Admiral Duguay-Trouin was becalmed in a fog off Spitzbergen for nine days. Because the sailors were unable to see the sun, the only idea they had of the time came from their hourglasses. When the sun reappeared, it was discovered that the sandglasses were off by eleven hours.

EUSTACHIO PORCELLOTTI
COSTRUITO A FIRENZE
L'ANNO 1883
M.FCH.LAB.
337

Model of Galileo's escapement of 1641, constructed 1883

Galileo Galilei, the Italian scholar and scientist, is generally credited with the invention of the pendulum. The idea came to him in the cathedral at Pisa, while he watched a large lamp swinging from the vaulted arch of the roof. Using his pulse as a timekeeper, Galileo measured how long it took for the lamp to swing to and fro, and observed that this time was the same regardless of the size of the swings.

Although he made this discovery in 1581 (some authorities claim 1583), when he was only a young man, Galileo does not appear to have thought of applying it to timekeeping until 1641, the year before his death. He was then blind and gave instructions to his son Vincenzio for a design which would incorporate his pendulum and an escapement. It was evidently never completed, because at the death of Vincenzio's widow in 1668, the inventory of her possessions contained the entry "an unfinished iron clock with a pendulum first invented by Galileo."

This model of Galileo's escapement and pendulum was made in 1883. The escape wheel has twelve teeth which are locked by a hinged lever, or *detent*, and twelve pins which bear against an arm controlled by the pendulum. As the pendulum reaches the end of its swing to the left, the detent is raised and the escape wheel unlocked. It moves until one of the pins engages with the second curved arm carried by the pendulum. As the pendulum swings to the right, this arm is freed from the pin and at the same time the detent is released again to lock the wheel in a new position. This type of escapement, the pinwheel, did not come into general use for more than a century; had Vincenzio ever completed Galileo's clock, it probably would have kept excellent time.

Lantern clocks (front view)
Left: Jeffrey Bayley, 1653;
right: Thomas Knifton, 1645

Lantern clocks were among the very earliest domestic clocks used in England, the first appearing in about 1620. Both of these clocks were made in London: the one on the left by Jeffrey Bayley, "At the Turn Stile in Holburne" in 1653, the one on the right by Thomas Knifton, in "Lothbury, London" in 1645.

Clocks of this sort were made to hang on a hook on the wall, and were driven by weights (not shown here). They were of brass, sometimes decorated with silvered hour rings, as the Bayley clock is. They sounded the hours, and later examples also had alarm mechanisms. Rarely did they have more than one hand, although this simple hour hand, in the form of a pointer, became more elaborate the later the clock. Lantern clocks with two hands are either very late, or else they have been "modernized," often by Victorian clockmakers who had little sympathy with seventeenth-century craftsmanship.

The escapement mechanism of the earlier clocks consisted of a single-spoke iron balance wheel; this revolved through an arc of about 240°, was stopped by the stroke striking against a pin, and then bounced back until it struck the opposite pin. The balance wheel oscillated with a period fixed by the arc of its swing, and was thus the mechanical equivalent of a pendulum. Both of the examples shown here have a balance wheel escapement; later clocks were fitted with pendulums at the back.These swung from side to side and appeared briefly at the limit of each swing, thereby earning the name "bob" pendulums.

The fretwork above the face on lantern clocks can often provide clues about their age and origin; the dolphin frets shown here were very common on English models. In addition, the Bayley clock has the original owner's initials, T.R., and the date, 1653, engraved on the fret—a rather unusual feature.

Jeffrey Bayley at the Turne Stile
In Holburne Londini Fecit

Lantern clock by Edward Webb of Chewstoke, 1688 (rear view)

The back view of this clock, made by Edward Webb of Chewstoke in 1688, is typical of later lantern clocks. Unlike the two earlier clocks shown above, its timekeeping is regulated by a small pendulum rather than balance wheels. The pendulum (here locked in position for storage) normally swings to and fro through a large angle, appearing alternately on either side of the case. It is attached directly to an arbor (the clock-maker's term for an axle), which runs forward above a horizontal crown wheel. A small lever or pallet attached to the arbor engages with the teeth of the crown wheel, producing intermittent motion and a simple form of escapement. Clocks fitted with pendulums were not particularly accurate, although they may have been better timekeepers than the balance-wheel clocks that preceded them.

The clock also shows the locking wheel, or locking plate, which controls the striking; this is the circular iron wheel at the back of the clock with notches cut into its circumference at graduated intervals. When the lifting piece is locked into one of the notches—as it is here—the clock is silent. But when the hour is reached, the locking piece is lifted, allowing the striker to hit the bell. The number of times the bell is hit depends on how long the locking piece is held up until it again falls into a notch. Thus the notches are cut with increasing spaces, the greatest space corresponding to the strike of twelve, the shortest to a strike of one. Locking-wheel striking was used on English lantern clocks as early as 1625, but went out of fashion in the late 1670s when Thomas Tompion began to produce clocks with the new rack strike.

Underneath the clock can be seen the rope which would have been attached to weights to drive the clock. In use, the clock would simply have been hung from a hook on the wall so that the ropes could pass out of holes in the base plate. The fret on this clock is a lion and unicorn supporting a shield and crown.

Royal
Observatory.

This "watch machine," made by the London clockmaker Larcum Kendall in 1774, accompanied Captain Cook on his third voyage, the attempt to find a northeast or a northwest passage from the Pacific into the Atlantic. It was the third such watch Kendall had made, following in the footsteps of the great clockmaker John Harrison, who was the first to design a clock which could be taken to sea and survive rough weather while still keeping good time.

Accurate timekeeping at sea was of enormous importance because it enabled sailors to estimate their longitude with reasonable precision. They could determine local noontime by watching the sun and measuring its position accurately with an octant or sextant to record when it reached its highest position in the sky. If they also had with them a clock set to Greenwich time, by the difference between the sun and the clock it was possible to work out exactly how far east or west they were from Greenwich. So great was the importance of this calculation that the Board of Longitude offered a prize of £20,000 to any man who could devise a way of determining longitude at sea to within ½°, or thirty nautical miles. John Harrison won the largest portion of that prize, though not without argument; and then, to make sure that his methods could be applied by other clockmakers, Larcum Kendall was commissioned to copy Harrison's fourth chronometer. The result was his first marine timekeeper, which also traveled with Cook in 1776. The third chronometer, the one shown here, was a simplified version which did not, in fact, keep time as well as the first.

The third chronometer has a verge escapement, a steel balance wheel, and a special bimetallic spiral compensation curb which compensates for changes of temperature. It has three faces, one for hours, one for minutes, and one for seconds. After its return from the Pacific with Cook's expedition (Cook himself did not return; he was murdered on the island of Hawaii), this chronometer was sent in 1791 to the northwest coast of the United States with Captain George Vancouver on the new *Discovery*.

The practice of navigation at sea was revolutionized by the introduction of the octant, or reflecting quadrant, in the 1730s. It was devised by John Hadley, a student of astronomy who had spent some time improving the design of reflecting telescopes. His reflecting quadrant was announced to a meeting of the Royal Society—of which Hadley was then Vice President—early in 1731. Quite independently, Thomas Godfrey of Philadelphia had made and tested a reflecting sea quadrant of a similar design in 1730. But news of Godfrey's instrument did not reach London until 1732, and so it is the Englishman who is almost universally credited with the invention.

The essential feature of both Hadley's and Godfrey's designs was the observation of the sun, or a star, by double reflection. There is a fixed or horizon mirror on the left-hand limb which is half silvered and half clear glass. Holding the instrument vertical, the navigator views through the peephole sight on the right-hand limb and sees the horizon through the clear glass. At the same time, he moves the brass index arm until the image of the sun is also seen, reflected from the index mirror and the horizon mirror. From the calibrated scale on the arc of the instrument it is now possible to read off the elevation of the sun above the horizon. Because of the double reflection, the frame of an instrument measuring up to 90° need only subtend 45°, or an eighth of a circle: hence the instrument became known as an octant. The octant was much more convenient and accurate in use than earlier devices, and made it possible to measure latitude with certainty for the first time.

This octant is signed "Egerton Smith Liverpool," but was probably made in London in the middle of the eighteenth century. Smith was a printer, lecturer on science, and dealer in mathematical instruments.

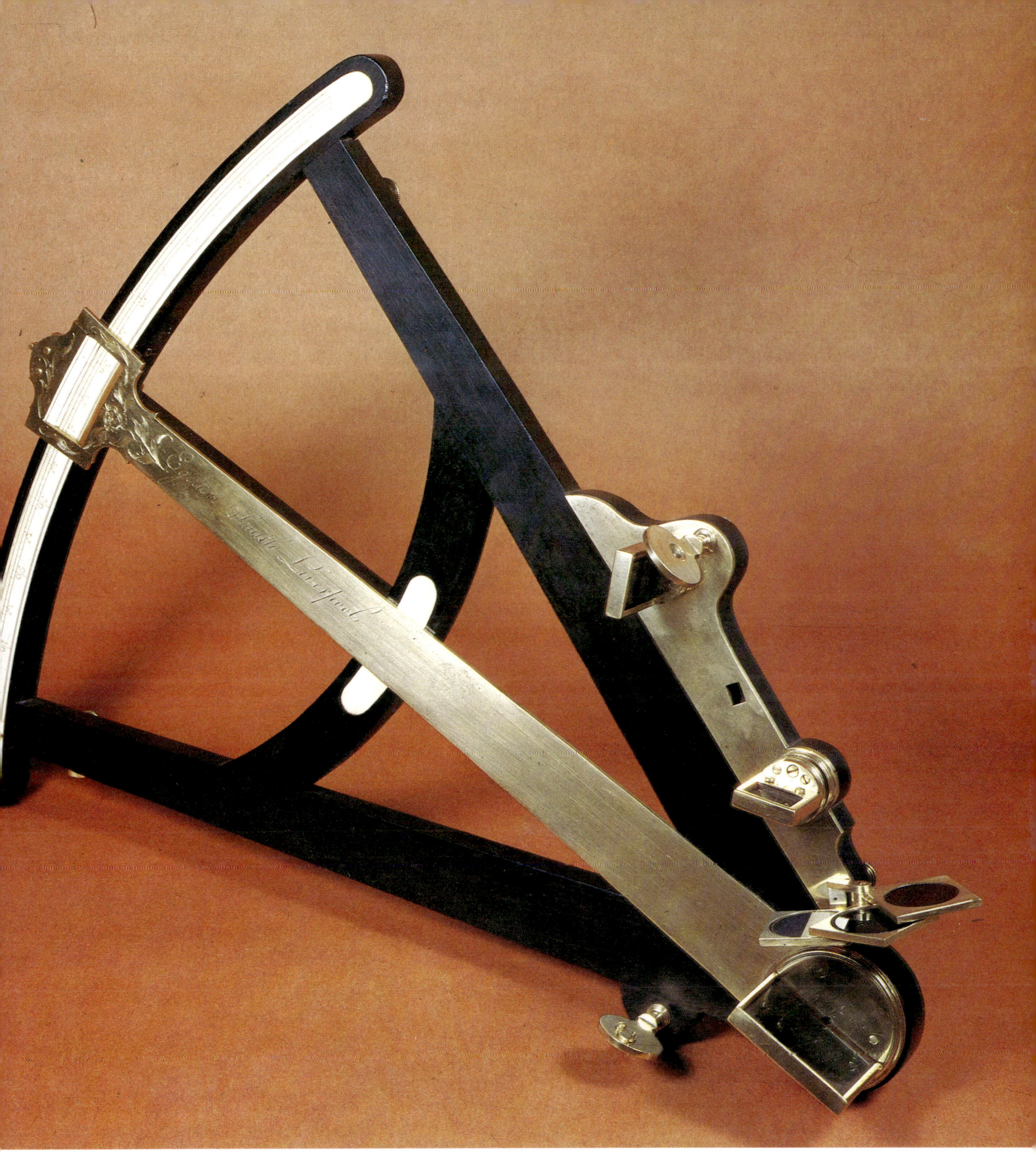
Liverpool

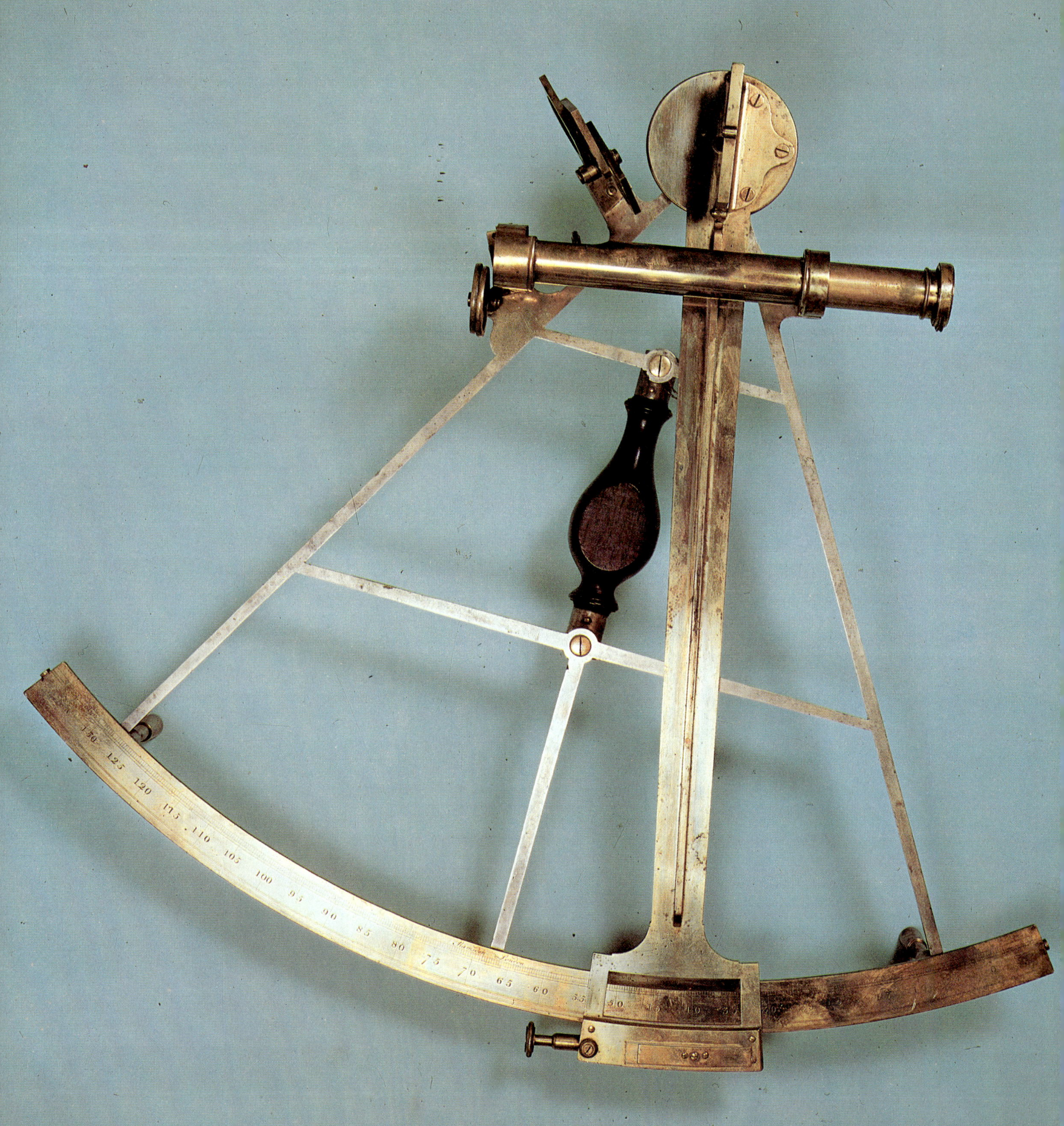

Hadley's reflecting quadrant of 1731 was followed in 1757 by a more accurate instrument based on the same principles: the sextant. It was introduced by Captain John Campbell, and met the needs of scientists and the more scientific navigators, including Captain Cook, for an instrument able to measure larger angles, with greater precision, than could the simpler and less expensive octant. The sextant shown here was made by Jesse Ramsden of London, one of the best instrument-makers of the day. From the late 1770s on, Ramsden assigned serial numbers to all his sextants; this is number 842, dating from 1785. It is fitted with a telescope, several filters to prevent glare when looking directly at the sun, and a scale somewhat longer than that of the octant. A sextant will read angles up to 120°, although because of double reflection the arc of the instrument itself is only half that—60°, or a sixth of a circle, which accounts for its name.

One major use of the sextant to the navigators of the eighteenth century was to measure lunar distances. To an observer on the earth, the moon appears to move across the heavens at an appreciable speed; in fact, it moves through an angle of 13° every twenty-four hours. At any given instant, therefore, the moon is at a certain fixed distance from any star, including the sun. These distances were tabulated in the late 1750s by Professor Tobias Mayer of Göttingen, and by measuring the distances and consulting the tables it is possible to work out the precise time at, say, Greenwich. By comparing this with local time, taken from the movement of the sun, a navigator could work out his longitude within ½°, or thirty nautical miles, a target of precision set by the Board of Longitude. To achieve this standard, it was necessary to have an instrument that could measure "lunar distances" within ten seconds of arc, a much greater accuracy than the octant could manage. The sextant was the answer.

The first sextants, made in the 1750s and 1760s, were substantial instruments, with a radius of fifteen inches or more. They had to be this large in order to produce a long scale, which could be graduated more accurately than a short scale with the techniques then available. A long scale spaces its divisions further apart than a short one does, and is therefore easier to make.

In the 1770s, however, Jesse Ramsden introduced new machines that revolutionized the graduation of both circular arcs and linear scales. His "dividing engine" for circular arcs consisted of a large bronze wheel almost 3½ feet in diameter, mounted horizontally on a steel shaft. The edge of the wheel had 2160 helical teeth cut into it, each one corresponding to an angular distance of 10 minutes of arc. The scale to be graduated was bolted to this wheel, which could be turned through precise angular distances by means of a gear engaging with the helical teeth. Thus the scale would be scribed, moved through ten minutes of arc, scribed again, and so on. In this way a scale could be graduated to an accuracy previously unobtainable.

The machine aroused considerable interest, and the Board of Longitude in London entered into an agreement with Ramsden. He was to be paid £615 on condition that he publish an account of his dividing engines and agree to divide sextants and octants for the trade at six shillings per sextant and three shillings per octant. So accurate were the scales produced by Ramsden's dividing engines that it became possible to make sextants much smaller without losing accuracy.

This pocket sextant was made by John M. Eckling of Vienna. Fashioned in brass with a silver arc, at 4½ inches radius it is a very much smaller device than Campbell's first sextant. Today's sextants have a radius of no more than 6 inches, yet they are very accurate instruments. Smaller sextants have the additional advantage of increased manageability—a great asset in rough seas.

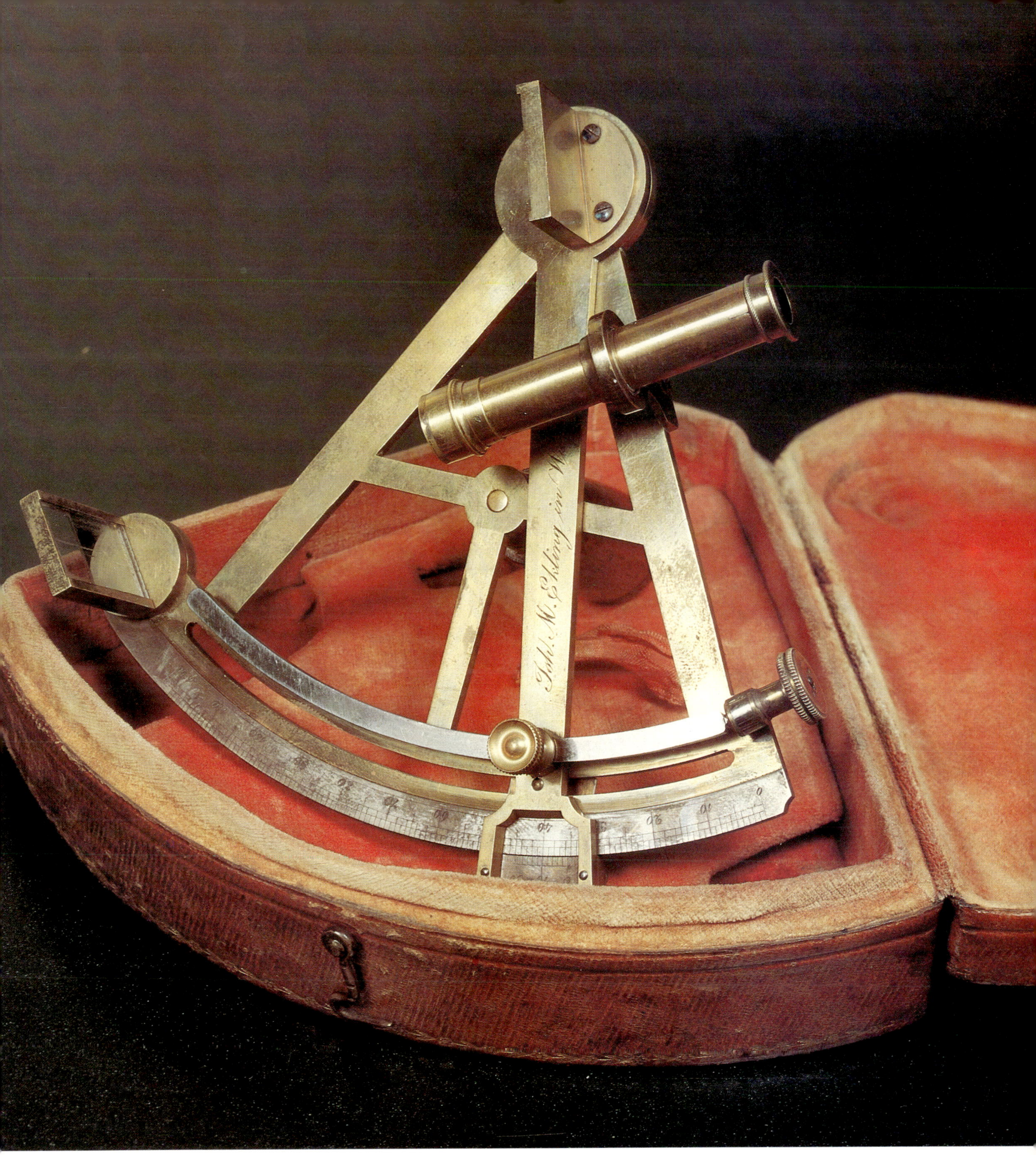

The traverse board was used as an aide-memoire by navigators as early as the sixteenth century to record, during each watch, how far and in what direction their ship had been heading. It consists of a simple wooden board, equipped with pegs which can be inserted into a series of holes. The upper, circular portion of the board is marked out with the thirty-two points of the compass; from the center radiate thirty-two sets of holes with eight holes in each set. A set of eight pegs (some are missing from the example shown) is attached with cords to the center of the circle.

At the end of each half-hour of the watch, the officer of the watch would take a peg and stick it into the hole on the compass bearing on which the ship had run during that half-hour. The first half-hour would be represented by the first hole cut from the center, and so on. At the end of the four-hour watch, all the holes would be filled with pegs, whose positions would give a good idea of the direction sailed during the watch. To estimate the distance sailed, the lower four rows were used, one for each hour of the watch; an estimate of the average speed achieved during that hour was made and the peg was inserted into the corresponding hole. At the end of the watch, the officer of the watch would transfer the information recorded on the traverse board onto a slate, or perhaps onto paper, and at the end of the day the captain would use this recorded information to write up his log.

Traverse boards provided a simple and relatively foolproof method of recording which could be used even in the foulest weather. They were adopted only by northern navigators—no Mediterranean example has ever been found—and continued in use up to the beginning of the twentieth century. This particular board dates from about 1800 and was found in 1844 on the Scottish island of Barra. Though its detailed history is unknown, the representation of East by the letter O indicates that it must have been used by Scandinavian, Dutch, or German sailors.

Before the discovery of electromagnetism, the only way to make an iron bar magnetic was to rub it with a lodestone, a naturally occurring mineral called magnetite. Chemically, this mineral is an oxide of iron (Fe_3O_4), with powerful magnetic properties. Who first discovered that a bar of iron rubbed with a lodestone would tend to point toward the north—and so invented the compass—remains a subject of controversy. The chances are that it was a Chinese living in the second or third millennium B.C.

The lodestones which made the apparent sorcery of the compass possible were naturally regarded with awe, and acquired an almost supernatural reputation. According to sixteenth- and seventeenth-century writers, the very best stones came from China and Bengal. They were large and heavy, and it was said that they could lift twenty times their own weight in iron. Eagerly sought after, they were commonly sold for their weight in silver. Good stones, such as this brass-bound specimen, which dates from the eighteenth century, were bluish in color. Reddish ones, of slightly lesser quality, were found along the Red Sea coast. Lodestones were also found in Asia Minor, Macedonia, Spain, Denmark, Germany, Siberia, and in the English Lake District.

As an item of natural magic, the lodestone attracted considerable interest. In 1756 the Countess of Westmoreland presented a large piece of lodestone to the Ashmolean Museum in Oxford. Cased with a brass coronet and complete with pole pieces, it weighed 171 pounds and could hold a load of 163 pounds. The example seen here, less spectacular but more practical, has been cased and shaped to be used as the essential tool of a compass-maker. To magnetize steel compass needles, all he had to do was rub them periodically against the stone.

In the Middle Ages, lodestones were widely believed to possess magical properties. If one were placed on a woman's head while she slept, for example, she would confess her adulteries. A lodestone held in the hand would cure gout, or give its possessor the power of eloquence. It was also believed that the lodestone's power coud be counteracted and its virtue destroyed by contact with garlic or diamonds. In 1669 John Seller, a nautical instrument-maker living in Wapping, London who served as hydrographer to Charles II, put this belief to the test: "There is a common received Opinion, [he wrote] that if the *Load-stone* be rubbed over with Garlic or Onions, that it will obstruct the Virtue there-of; or if a Knife being touched upon the *Load-stone* and afterwards cut an Onion or Garlic therewith it will immediately lose its Virtue. This conceit hath also been countenanced by the Ancients, but if you are pleased to make a trial of it, you will find it but a meer fallacy. It is also false, that the *Diamond* doth hinder the Virtue of the Load-stone while it is near it."

Lodestones were often elegantly mounted and provided with handles and carrying cases. Mineralogically, they are naturally occurring iron ores which picked up their magnetism from the magnetic field of the earth as they cooled. The lodestone on the left is shown with a mass of iron filings, which are strongly attracted to the two iron poles surrounding the stone. The other specimen has a pierced brass case decorated and engraved in a style characteristic of late eighteenth-century Russia.

By the late sixteenth century, the mariner's compass had evolved into an instrument not very different from the compass of today. This example, made in Italy about 1570, is probably the oldest surviving compass in Europe. It was constructed in the way described by the Spanish writer Martin Cortes in his work *The Art of Navigation*, published in 1596.

"Take a piece of paper and cut it into a circle of diameter equal to the span of a man's hand," Cortes advised. "Mark on it the 32 points of the compass, distinguishing North by a fleur-de-lys." The use of the fleur-de-lys, it is believed, evolved from the initial letter of the Italian word for north, "tramontana." In Medieval times, the T became elaborately decorated and evolved into the fleur-de-lys—perhaps, some have speculated, through the influence of the French fishermen of Aquitaine.

Cortes continued: "Take a wire of iron or steel of the largeness of a great pin and fix it to the underside of the card and touch it with a lodestone." To the center of the card was attached a cap or bearing, normally made of brass, and then the whole card, or "fly," was balanced on a pivot, about half the diameter of the card in height. The compass was mounted in a box made of non-magnetic materials—usually wood, but ivory in this case—and often covered with a glass lid to protect the card and the needle from the weather. This box was then commonly mounted inside a second box in which it was held by two gimbals, rings of brass, which pivoted so that the compass would remain horizontal even if the ship were rolling and pitching. The compass shown here is a very early example of this kind of mounting, and the fly, mounted on a pivot according to Cortes's description, is decorated at both east and north. It is not known who used this compass, but similar instruments enabled the explorers of the sixteenth century to make their epic journeys.

The compass changed very little in the two centuries between 1550 and 1750. In 1616 William Barlow complained that "the compass needle, being the most admirable and useful instrument of the whole world, is both amongst ours and other nations for the most part, so bungerly and absurdly contrived." And in 1754 John Robertson, a Fellow of the Royal Society and teacher of mathematics at Christ's Hospital School, wrote a highly critical description of the compasses then in use. His book, *Elements of Navigation*, was published at almost the same time as the compass shown here was made by William Farmer of London.

The problems identified by Robertson included the use of soft iron wire for the compass needle and iron nails to hold together the wooden cases; iron's metallic properties introduced errors into the compass. "Scarce one sea compass in ten is fit for the use for which it is made; and this has arose from their being fabricated by unskilful and ignorant workmen for the wholesale dealers in the shipping way; who generally pay no more regard to the construction of this instrument, whereupon the success of the voyage and the lives of the men in great measure depend, than they do to any indifferent thing of the same price."

Farmer's compass is signed on the engraved and hand-painted card: "Made by William Farmer near the Limekiln, Horsely-down." The card is mounted on a brass pivot set on the base of the turned-wood bowl—a traditional if basic design that persisted well into the nineteenth century. The base can be readily detached to allow the needle to be retouched with a lodestone. The compass card is attractive, with a large fleur-de-lys north point. The other seven cardinal directions of the compass rose are indicated by allegorical representations of the sun, moon, and planets.

Made by William Palmer near the Limekiln Horsley down

S
SE
E
NE

This compass, dating from about 1770, is a good example of the improvements made to the compass in the second half of the eighteenth century by Dr. Gowin Knight, F. R. S. Compasses of this design were known as azimuth compasses because they could be used to take bearings, and a compass much like this one would have been used by Captain Cook on his first voyages of discovery.

This compass card is signed "Invented and Made by R. Rust and R. Eyre in ye Minories London." The design differs only marginally from that patented by Knight in 1776 and made under license by George Adams of Fleet Street. Dr. Knight became interested in compasses in 1749 after seeing one that had been struck by lightning and ruined. The lightning had magnetized the iron nails and wires used in the construction of the compass box, which caused wild errors in the directional indication. So poor were the general run of compasses then available that Knight decided to redesign them. He tested various types of compass needles and found that the best were perfectly straight needles, properly tempered, rather than the diamond-shaped wire arrangements then in common use. He used thinner paper for the card, supporting it around its circumference by a brass ring. He also studied bearings and found that the best could be made from a small piece of agate set into ivory. The best pivots, he said, were common sewing needles; he would not trust the compass-makers to produce them, but preferred to get them directly from the needle-makers, who understood their manufacture and made a dependable product.

This azimuth compass is equipped with sights, so that it can be used to take bearings. It is mounted on gimbals, so that it stays horizontal; its major fault, according to the navigators of the day, was that it tended to be unsteady in rough weather.

Thalen-Tiberg magnetometer
with Dahlblom Sine Arm, c. 1890

Very early in the history of navigation, it was realized that a compass may not always tell the truth. In particular, it may deviate when it is used close to buried metallic ore, which deflects its needle. In the second half of the nineteenth century, this observation began to be put to use in the form of geomagnetic prospecting, using such deviations to plot the position of useful deposits of ore.

This instrument, the Thalen-Tiberg magnetometer, was made by the Berg Company of Stockholm about 1890. It can be used for measuring either the horizontal or the vertical intensity of the earth's magnetic field; here it is set up for a vertical measurement. At the end of the shorter of the two arms it carries a deflecting magnet, in the same vertical plane as the compass needle, which it deflects. The degree to which the main compass needle is deflected by the small magnet on the end of the arm gives a measure of the magnetic field at that place on the earth's surface. The arm can be pointed in any direction by means of a foresight and backsight, and it carries a millimeter scale to show the distance of the deflecting magnet from the compass.

The second, longer, arm is called a Dahlblom Sine Arm and provides an alternative method of measuring the horizontal component of the earth's field. In this case the deflecting magnet is slid along the arm to produce a constant deviation in the compass, and a scale along the arm enables the horizontal intensity of the field to be read off directly.

The first compound microscopes appeared during the seventeenth century, although single lenses had long been used as magnifying glasses. The idea of fitting two such lenses together and mounting them in a tube to produce greater magnifications may have been Galileo's; but invention of the microscope itself is difficult to attribute unequivocally to any single person. By the 1670s, Italian makers were producing microscopes like this handsome model, with its tooled leather cover. The cardboard body is made in two parts which slide inside one another. Focusing is done by a coarse screw adjustment, which raises and lowers the microscope shaft, and a second screw which adjusts the eyepiece. When not in use, the eyepiece is protected by a screw-on wooden cover or cap.

Such microscopes opened up a new world to the scientists of the day. Not only could they see familiar objects greatly enlarged, but, more important, they could also study objects never before seen by man. One famous observation made in 1663 by Robert Hooke, Curator of Experiments at the Royal Society, was of the structure of a thin slice of cork. It was, he noted, made up of a fine pattern of small holes, which he called cells. Later, these cells were shown to be the remnants of structures which in life are filled with fluid and form the basis of all living systems; the name cell, first used by Hooke, is still used to describe them.

Microscopes became something of a passion for the educated man of the day; Samuel Pepys bought one in 1664 for what he described as "a great price"—£5 10s. Pepys also bought Hooke's book, *Micrographia*, published the following year, and sat up until 2:00 A.M. reading it. It was, he wrote, "the most ingenious book I ever read in my life."

Since the late 1920s, this beautiful microscope has been attributed to the personal collection of Robert Hooke (1635–1703), the pioneer of microscopical investigation. As such, it has made appearances in many reference works. However, recent research has shown that the claim is not true.

There is no doubt that the instrument is original. It was made in London in the 1680s by the renowned optician Christopher Cock, and was part of the scientific apparatus owned by George III. In 1841, the collection was dispersed and this microscope, together with many other instruments, was presented to King's College, London, for use in teaching a "General Course of Experimental Philosophy." In the 1880s all the microscopes were given to Sir Frank Crisp, a noted microscopist, with the implicit understanding that he would bequeath the whole of his large collection of microscopes to the nation on his death. Alas, Crisp's will contained no such provision, and his collection was auctioned off in 1925.

Thomas Court, a dealer in early instruments, bought the microscope and subsequently presented it to the Science Museum, and it was he who originally declared that the microscope was Hooke's own. To be sure, the inventory of Crisp's collection described it as "Dr. Hooke's Microscope," but this did not mean quite what it seems to. In the eighteenth century it was common to use such a term to describe a microscope merely made to the inventor's design. If the compiler of an eighteenth-century inventory knew the maker, he might write, for example, "Jobolot's Microscope by Costa of Bordeaux." Court was misled by the description and was further confused by the maker, who claimed that the gold tooling was identical with that on a microscope signed by Cock. Some remarkable detective work has shown not only that Court's particular claim was false but also that optical instrument-makers brought in tubes already made and decorated from specialist tube-makers. Therefore, the design of the motifs on the tubes gives no reliable guide to the identity of the optician who ground the lenses and fitted up the complete instrument.

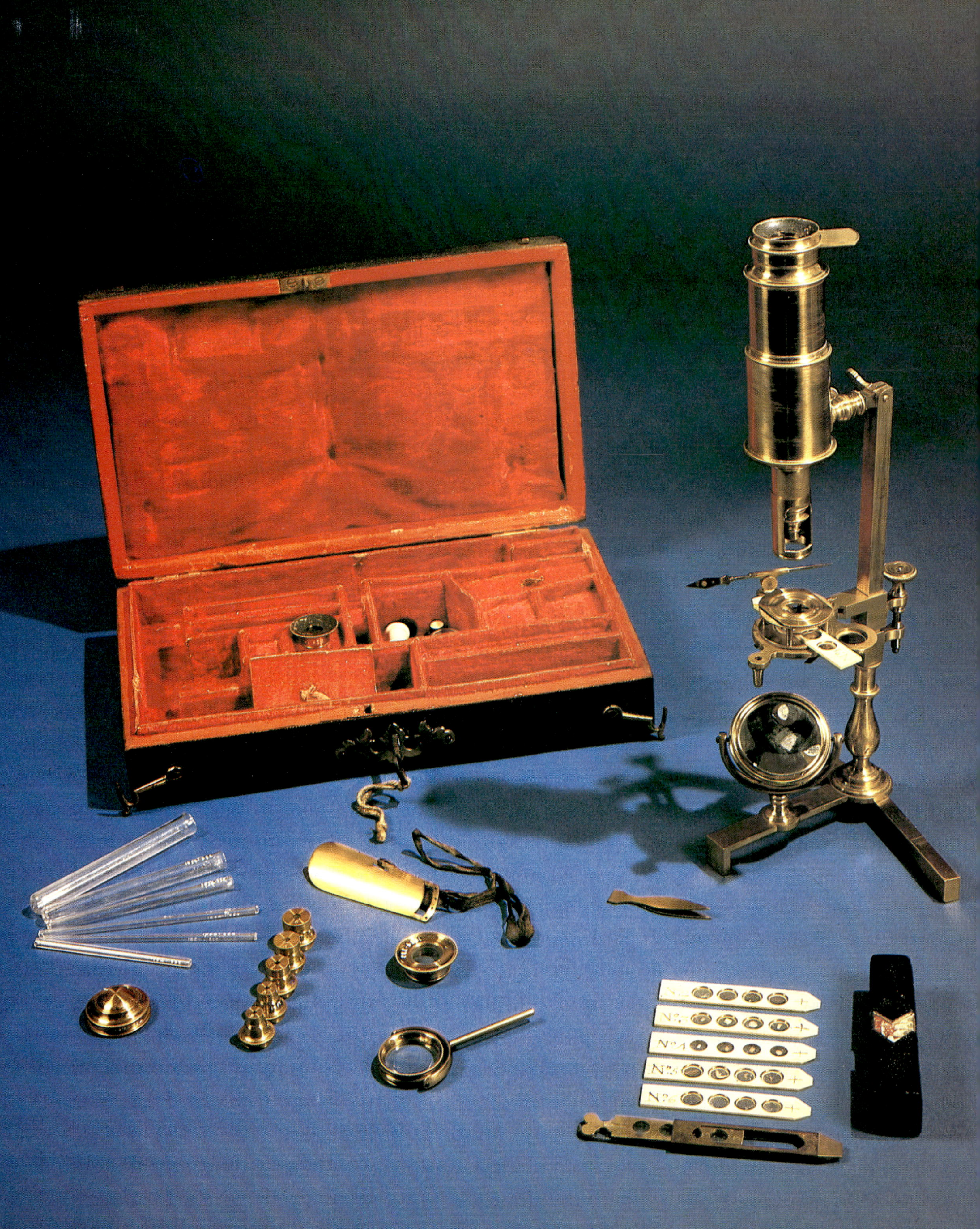

In 1767 Joseph Priestley paid £5 15s. 6d. to the instrument-maker Benjamin Martin of Fleet Street (then a center for instrument-makers) for this microscope, described as a "Martin New Universal Compound Microscope in a Shagreen Case with apparatus complete."

Priestley was one of the founding fathers of chemistry. He discovered a number of unknown gases, including carbon dioxide, ammonia, and hydrogen chloride—and, most important of all, oxygen. Priestley, who was a believer in the then-current phlogiston theory of combustion, called this new gas "dephlogisticated air"; it was left to Antoine Lavoisier to name it oxygen a few years later. Using this microscope, Priestley showed that air impaired by putrefaction, the breathing of animals, or the burning of candles was restored by green matter which grew in water. Thus he demonstrated that green plants absorb carbon dioxide and give out oxygen.

Priestley was a militant nonconformist in religion, a supporter of the American colonists in their revolt against the British, and a sympathizer with the French revolutionaries. He was regarded with some suspicion by his neighbors in Birmingham, where he had settled in 1780, and in 1791 a mob burned down his house, laboratory, and the New Meeting House where he was employed as minister. He escaped to London, and in 1794 emigrated to the United States, where he was welcomed. There he spent the last ten years of his life peacefully, dying in Northumberland, Pennsylvania, in 1804. It was appropriate that Priestley should end his life in America, for it was Benjamin Franklin, whom he had met in London in 1765, who had first encouraged his interest in science.

In the 1780s George Adams made two of history's most ornate and curious microscopes. Rarely has fine craftsmanship been put to such eccentric use. The microscope shown here, apparently constructed for the Prince Regent, is made of brass and steel and covered with beaten silver. Allegorical figures grapple around the eyepiece, while classical urns and mermaids festoon the pedestal. It is magnificent—but does it work? It is too tall to use easily on a table, too short for the floor, and the lenses are not very conveniently placed. But it must have looked superb standing in a corner of some drawing room at Carlton House, where the Prince Regent lived, and it remained in the possession of the Royal Family until George VI gave it to the Science Museum. Its twin, made for George III, is now in Oxford.

Aside from its unnecessary ornateness, this microscope does show some of the features of Adams's more down-to-earth instruments. In 1746 he had introduced his "New Universal Single Microscope," which had a rotating arrangement of single lenses of different powers; the lenses could be brought over the object to be viewed simply by turning the plate that held them. A similar arrangement can be seen on the silver microscope: a rotating stage carrying different lenses lies just beneath the figures. The silver microscope also has a pair of mirrors to throw light onto the two stages that hold the slides—a device typical of the period.

Adams was a remarkable craftsman who, together with his son (also called George), made a huge range of different instruments. These included compasses, micrometers, pantographs, spectacles, theodolites, prisms, barometers, telescopes, microscopes, and zogroscopes (instruments for viewing perspective prints). Father and son both held the appointment of Mathematical Instrument-Maker to the King. In the earlier years their premises were in Racquet Court, just off London's Fleet Street. In this court, which was once part of the Fleet Prison, the game of racquets was invented by inmates seeking exercise.

Facsimile of Newton's reflecting telescope, 1671

Isaac Newton made his first reflecting telescope in 1668 and a second, at the request of the Royal Society, in 1671. This is a copy of that treasured instrument, one of the relics of England's most famous man of science.

Newton had turned to reflecting telescopes after concluding that it was impossible to produce a refracting telescope free of the serious problem of chromatic aberration. When light passes through a simple lens, just as when it passes through a prism, the different colors that comprise white light tend to be bent to varying degrees. The result is that telescopes using simple lenses always showed stars with a distracting ring of color around them.

Newton's contemporaries believed that the problem would be solved if lenses could be ground to a non-spherical curvature. But Newton knew that mirrors do not suffer from chromatic aberration, and he determined to make a telescope based on reflection rather than refraction.

Others had designed reflectors from the 1620s on, but Newton's, built during the summer of 1668, was the first successful working model. It consisted of two mirrors. One, with a concave surface, was placed at the lower end of the telescope; light falling on it was concentrated and thrown back up the tube to fall on the second, flat mirror. This mirror was placed at an angle so that it reflected the light rays out through the small eyepiece near the upper end of the telescope. This system eliminated any chromatic aberration, and had a second advantage as well: since the mirrors did not absorb any light, as lenses do, the telescope was capable of detecting many more stars than earlier models could.

Newton's second reflector, made in 1671 for the Royal Society, was his passport into the scientific circles of the time. The day it was exhibited, he was elected a Fellow of the Society. To his contemporaries, the most significant feature of Newton's telescope was not the absence of color, but its small size. This reflector, only six inches long, performed as well as a refractor four feet long.

This telescope and its equatorial mounting were made by the remarkable Jesse Ramsden (1735–1800), a man who brought new standards of precision to instrument-making. He patented this design in 1775. Made of brass, the telescope is of the refracting variety, and the whole apparatus stands about two feet high. An equatorial mounting was used by Tycho Brahe in the late sixteenth century, but this particular style was first made in or before 1737 by Jonathon Sisson of London.

Although the mounting arrangements look needlessly complex, each element serves an important purpose. Viewed from the earth, all the stars in the heavens appear to rotate about a central point, the North Star. In fact it is the earth which is rotating, and the North Star remains still only because it happens to lie on the earth's axis of rotation. To follow a star over a lengthy period of observation, it is helpful to have a telescope which will turn about the same axis as the earth. Then, by slowly rotating the telescope about this axis, it is possible to keep the star in the center of the field of view all the time. A plane at right angles to this axis is of course parallel to the equator, and is called the equatorial plane; and a telescope so mounted is known as an equatorial telescope.

With a portable equatorial such as this one, a further complication is introduced. The inclination of the polar axis depends on the latitude from which the observation is being made, so the slope of the polar axis must be adjustable. In addition, to point the telescope to different heights above and below the equator it must be able to be turned about a second axis (known as the declination axis), set at right angles to the polar axis. Ramsden's mounting provides both these axes, and was popular with amateurs because it allowed the instrument to be used not only as an equatorial but also as a transit and and altazimuth instrument for astronomy, and as a level and a theodolite for surveying. This versatility gave rise to the popular name of "portable observatory."

John Russell (1745–1806) was a portrait painter greatly esteemed by his contemporaries, who thought him as good as Sir Joshua Reynolds. Despite his abilities, Russell, an ardent Methodist, did not receive formal recognition until he was in his mid-forties. His election to the Royal Academy in 1788 was followed by an appointment as Painter to the King and Prince of Wales in 1790. As a painter, Russell was dissatisfied with the existing lunar charts, which he considered to be aesthetically unattractive representations of a beautiful celestial object, and contact with leading astronomers convinced him that contemporary maps of the moon were also technically inaccurate. In 1785 he began a systematic series of observations using a telescope fitted with a micrometer, which could measure the size of objects on the lunar surface.

The study of lunar geography is called selenography, from the Greek word for the moon; Russell called his twelve-inch-diameter moon globe a selenographia. The globe, printed from plates he had engraved, and its complex stand were patented by Russell in 1796 and put on sale in June 1797. The globe does not, of course, show the whole of the moon's surface, since the moon always presents the same face to the earth. Its dark side remained hidden from man until satellites circled it in the 1960s.

However, the globe does show rather more than half, since the moon appears to "wobble" when seen from the earth, allowing glimpses of the North and South Poles and around the edge of the eastern and western limbs. This wobble, known as a "libration," means that an earth-based observer can in fact see 4/7 of the lunar surface, while the rest remains permanently hidden. To demonstrate these librations, John Russell devised a support for his selenographia which permitted it to rock to and fro, and fitted to it a small model of the earth to show how its rotation affects an observer's view of the moon.

The oldest surviving terrestrial globe was made in 1492 on commission for the citizens of Nuremberg. But although a globe offers the possibility of perfectly accurate representation of continents and oceans, without any of the distortions which are inevitable when the earth's curved surface is represented on a flat map, early globe-makers were often more interested in ornamentation and aesthetics than accurate cartography. And although the globes were often taken to sea, they were probably too small to be of much help to a sailor. It is more likely that these globes, made by pasting strips of printed paper onto a sphere, were used to show gentleman adventurers on board the rough position of the vessel, and to teach midshipmen the rudiments of cartography and navigation.

This globe, some seventeen inches in diameter, was made in 1541 by the Flemish geographer Gerhard Mercator. Born in 1512 in Rupelmonde, in what is now Belgium, Mercator was a pupil of the renowned instrument-maker, geographer, and mathematician Gemma Frisius, who taught at the University of Louvain. Mercator engraved his first map in 1538 and the plates for this globe in 1541. It bears the legend and dedication: "Published by Gerhard Mercator of Rupelmonde under the patent of his Imperial Majesty for six years, at Louvain in the year 1541. Dedicated to the very distinguished Seigneur Nicholas Perrenot of Granvella, first counsellor of his Imperial Majesty."

Mercator is best remembered today for his projection, a solution to the problem of displaying the earth as a flat surface. In 1569 he published his masterwork, a map of the world "ad usum navigatum," which used his cylindrical projection. It is this projection which has produced the map of the world familiar to us all today.

An orrery is a working model of the solar system, showing the movement of the earth, moon, and planets around the sun. It gained its curious name by accident; the first (or actually the second) such machine ever made, by John Rowley of London in or before 1713, was commissioned by the fourth Earl of Orrery. Sir Richard Steele, the essayist, assuming that Rowley was the inventor of the machine, wrote that Rowley had named the machine an orrery in honor of his patron. In fact, Rowley had copied his machine from a design originated by George Graham, a leading instrument-maker of the day and a Fellow of the Royal Society, for Prince Eugène of Savoy. Properly speaking, Steele ought to have christened the invention a eugène, or perhaps a savoy; but he called it an orrery, and the name stuck.

The device shown here was made in London about 1800 by Edward Troughton. He and his brother John worked on Fleet Street and were famous as makers of astronomical and mathematical instruments. The first orrery showed only the motions of the sun, earth, and moon, but this one shows the planets Mercury and Venus as well. It is about twelve inches across, rather smaller than most other orreries. The central brass sphere is the sun, and the two white ivory balls circulating around it are Mercury and Venus. The Earth, modeled considerably out of scale, also revolves around the sun, and rotates on its axis. The moon, a small ivory ball, orbits around the earth.

The orrery can be used to demonstrate a number of astronomical phenomena. In this model, a metal arc suspended near the surface of the earth shows which parts of the earth are illuminated by the sun, and therefore where it is night and where day. A pointer moving around the horizontal ring indicates the date and the constellations of the sun. The moon is half covered with a black cap, which shows its phases. The whole apparatus is operated with a handle, which can be seen on the right.

In the mid-seventeenth century, accurate mapping by trigonometrical survey began. This method depends on only one measurement of distance: the baseline, from which are constructed a series of triangles. The angles of the triangles are measured and the lengths of their sides calculated by trigonometry. France was the first country to be mapped by such methods, and Britain followed. The first accurate surveys in England show that pre-existing country maps could be in error by as much as three miles in eighteen.

For the precise measurement of angles, a theodolite was used. This one was made by Benjamin Cole and is typical of the instruments in general use in the third quarter of the eighteenth century. It is decorated with the Royal Coat of Arms, since it originally formed part of King George III's collection, and the inscription reads "Cole, maker at ye ORRERY in Fleet Street, LONDON." It is, technically, an altazimuth theodolite, which means it can measure both altitude and azimuth, or both vertical and horizontal angles. The first such instrument appeared in Europe in the sixteenth century, but it was of very limited accuracy; it was simply equipped with open sights, through which the surveyor lined up the instrument on a distant "target." By the third decade of the eighteenth century, London's instrument-makers were beginning to use telescopic sights, incorporating crosshairs for precise alignment.

Cole's theodolite is equipped with two spirit levels, which are essential for getting the instrument horizontal before making a measurement, and a compass in the base. It has no leveling screws, presumably because these formed part of the base on which the instrument originally rested.

The abacus is one of the oldest, simplest, yet most powerful tools ever invented. Its origins are a mystery. The Greeks and Romans both used a counting board or table (*abakion, abacus*), arranged in columns on which counters were placed to represent numbers. The type of abacus shown here, in which the beads are fixed on the column and divided into two groups by a bar, came into use in the late Roman period. This example is Chinese, and dates from the middle of the nineteenth century.

With this type of abacus, beads in the upper portion, known as "heaven," count five units each, while those in the lower portion, "earth," count one. Calculations are carried out by moving the beads to the bar. For example, a wire which has one bead at the bar from heaven and one bead at the bar from earth represents the number 6 (5 + 1). The number stored on the abacus as it is shown here is therefore 1532786. (The two rods at the left-hand end of the abacus are not in use, since neither has a bead at the bar. Both represent 0.)

In Western Europe, the abacus began to disappear from use in the fifteenth century, probably because the Arabic numeral system and the availability of paper made written arithmetic easier. In countries which did not adopt the Arabic numeral system, though, the instrument survived and flourished. Until the pocket electronic calculator became popular in the early 1970s, the abacus was unrivaled as a portable, easy-to-use aid in calculation. It is a remarkably rapid device for addition and subtraction, though not so effective for multiplication and division. In a famous contest arranged by the U.S. Army in 1945, a junior official of the Japanese Ministry of Communications, using an abacus, easily defeated a U.S. Army finance clerk who used a mechanical office calculator. Today it is still in use in Japan, China, and the Soviet Union.

For over 200 years, the sector was one of the most important aids in calculation. Its origins can be traced to the proportional compass of the sixteenth century, and it reached its definitive form in 1598 in the hands of Galileo Galilei in Italy and Thomas Hood in England. The device consists of a hinged and graduated rule, used with a pair of dividers; computations are made according to the principle of similar triangles. The sector was particularly useful for working out proportions, for answering questions such as "50 is to 70 as 90 is to what?"—somewhat like a modern slide rule.

Sectors were produced with a wide variety of scales along the hinged arms: equal divisions, chords, trigonometrical and logarithmic functions, and so on. Specialized instruments were made for the use of architects, surveyors, navigators, and gunners. The sector shown here, made of ivory with a silver hinge, was the work of George Adams of London, and dates from the middle of the eighteenth century. Used for architectural calculations, the instrument enabled an architect to calculate the correct proportions for doors, columns, pedestals, and other decorative features for buildings based on the classical order. Thomas Carwitham, an architect and painter, was responsible for this sector's design, which was published in 1723 by the instrument-maker Thomas Heath. Interestingly, George Adams was one of Heath's pupils.

The sector was such a quick and useful calculating tool that it survived until well into the nineteenth century, despite competition from instruments such as the slide rule. Today hardly anybody but a historian of science would know how to use one.

Doric Flutes in Plano
Doric Upright Flutes
Flutes & Fill. of Columns in Plano
Frieze Heights of Entablatures of Doors
Heights of Doors and Entablatures
Projections Lesser Arches
Heights of Columns
Widths
Doors

The lid of this instrument bears the inscription: "Machina Cyclologica Trigonometrica . . . a Samuele Morlando inventa—Anno Salutis MDCLXIII." Its designer, Samuel Morland, was a prolific inventor of mechanical contrivances, though he did not perhaps invent quite as many as he claimed. His mechanical calculator of 1666, for example, was based on designs originated by T. L. Burrattini. But the instrument shown here, designed in 1663, does appear to have been Morland's idea.

This model, made of brass and partially silvered, was built in 1664 by Henry Sutton, an instrument-maker, and Samuel Knibb, a clock-maker, both of London. It can be used for simple multiplication and division, and also for any problem in plane geometry which could be solved by plotting on paper.

The trigonometer consists of a horizontal central scale, which can be moved up and down across the face of the instrument by means of a rack-and pinion operated by a capstan on the lower right-hand side of the instrument. The central part of the instrument is a toothed wheel just over eight inches across, which can be rotated by means of the other capstan. This wheel carries two further scales, one fixed to the wheel so that it rotates with it, the other either fixed to the wheel or aligned with the horizontal scale by means of a spring catch. Around the outside of the wheel is a series of scales which allow angles to be set. In use, the machine operates as the mechanical equivalent of the method of similar triangles used for multiplication and division. The sines, cosines, tangents, and other angles of any triangle can be read directly off the machine.

Morland appears to have been something of an opportunist. Toward the end of his life, in 1861, he was appointed Master of Mechanicks to Charles II; but the post held no salary, and Morland was a man whose expenditures characteristically exceeded his income. So in 1687 he married (for the fifth time) a woman he supposed to be an heiress. However, the lady, a Mistress Mary Aylif, turned out to be penniless.

Henricus Sutton
Londini
16
et Samuel Knibb
fecerunt
64

Millionen
Hunderttausender
Zehentausender
Tausender
Hunderter
Einheiten

J. Sauter (1723–1786), a German watchmaker, built this ornate and elegant calculating machine based on principles developed by the philosopher Blaise Pascal in 1642. The instrument consists of seven pairs of wheels, each wheel with the numbers 0 to 9 around its circumference. One pair represents units, one tens, one hundeds, and so on, up to the wheel on the extreme left, which represents millions. The machine is equipped with a stylus for turning a pointer; each of the front, or "setting," wheels has its own pointer, which can be set to any number between 0 and 9.

To add on a machine like this, the setting wheels are turned with the pointer. These wheels are geared to the rear set, the "result" wheels, so that any fraction of a turn of the setting wheels, from one tenth to nine tenths, advances the result wheels correspondingly by any number between one and nine. Thus simply by turning the setting wheels by the relevant amounts, the result wheels record the addition—except for one snag. When any column exceeds nine, some mechanical means must be devised for "carrying" a digit to the next column. When all the columns are standing at 9, this mechanism must be sensitive enough to propagate the "carries" right along the row, converting 999,999 into 1,000,0000.

This problem generally prevented machines like Sauter's from working well. Pascal, who made the first attempt at such a machine in 1642, used an ingenious ratchet arrangement which was gradually raised as the setting wheel was turned and finally fell as the number 9 was passed, giving the next wheel a fraction of a turn as it did so. But the difficulty of making such arrangements work smoothly, and the slowness of operation—no faster than pencil and paper—meant that calculating machines like this one had a limited usefulness.

In the early 1620s, William Oughtred invented the slide rule, a neat and handy device which used the logarithms of John Napier to mechanize the processes of multiplication and division. A simple slide rule has two scales arranged so that they can slide back and forth alongside each other. Each scale is logarithmic: that is, the numbers on it fall at a distance from the end which is proportional to their logarithms. In using the device to multiply, in effect two lengths are added together. This is equivalent to adding two logarithms, the process of multiplication.

The slide rule is a simple analog computer, a machine in which a physical property—in this case, length—is used to represent a number. The accuracy of a slide rule depends on the precision with which the divisions on the scale are made; and that, in turn, means that the longer a slide rule is, the more accurate its results can be. With this in mind, an American inventor, E. Thacher, patented the ingenious slide rule shown here. It consists of a cylindrical slide, which can be moved both in rotation and in and out longitudinally, and twenty triangular bars. Wrapped around the cylindrical slide are two complete logarithmic scales, whose total length is forty times greater than the length of the instrument itself. By moving the slide against the triangular bars, the normal operations of a slide rule can be carried out; but because the scales on the slide are much longer than they would be on a straight slide rule of the same dimensions, greater accuracy is possible.

Thacher patented his slide rule, which is capable of giving results accurate to four figures, in 1881. It was manufactured by the Keuffel and Esser Company in New York.

CALCULATING INSTRUMENT

Le Calcul instantané
ARITHMOGRAPHE TRONCET
SOUSTRACTION
ADDITION
PARIS
LIBRAIRIE LAROUSSE
19, rue Montparnasse, 19
Breveté en France (s.c.d.g.) et à l'Etranger.
Tables de l'arithmographe.

"Instantaneous calculation," promised the arithmographe of Troncet, patented in France in 1889 and produced by the Librairie Larousse in Paris. Whether the machine quite lived up to this billing is uncertain, but it could add and subtract and, through the use of printed tables which were supplied with the machine, also perform multiplication and division.

The arithmographe has seven vertical slots, each numbered from 0 to 9. Under each of the slots is a slider with twenty holes in it, going from 0 to 9 and back to 0 again. The upper half of each slider is black and the lower half white. The numbers on the sliders register in holes above and below the slots.

To use the instrument for addition, the special pointer provided with the machine is inserted into the hole opposite the number to be added and the slider pulled down all the way, until the pointer comes to rest at the bottom of the slot. This causes the number shown in the peephole to increase by whatever number has been "dialed." However, it is also necessary to "carry" 1 to the next column whenever the column being added exceeds 9. This is where the double scales on the slider come into play. If the number into which the pointer is inserted is in the black portion of the slider, the pointer can be slid upward and around the "hockey-stick" curve at the top of each slot. This operation automatically performs the carry by adding another digit to the next column.

In 1784 George Adams (the younger), mathematical instrument-maker to George III, expressed the view that "the science of Electricity is now generally acknowledged to be useful and important; and there is great reason to think, that at a future period it will be looked up to as the source from whence the principles of natural philosophy must be derived; its utility to man will not be inferior to its dignity as a science." How right he was.

The instrument that really brought electricity to the attention of scientists and the general public was the tribo-electric generator, a machine for generating electric charges by friction. The process had been known, but not understood, for centuries, and was brought to a new height in the middle of the eighteenth century by machines such as this one. Joseph Priestley, who wrote the first history of electricity in 1767, is reputed to have been the owner of this generator, which consists of a glass cylinder mounted on a horizontal axle, with a silk "apron" that lies across the top of the cylinder. A silk cushion rubs against the glass as the cylinder is turned by the handle. It was believed that the apron prevented the escape of the charge, which would be collected by the conductor, a hollow brass cylinder set on an insulated stand on the far side of the generator.

In 1749 Benjamin Franklin presented a machine almost identical to this to Yale College, and within a decade electrostatic machines had become all the rage. Franklin turned the phenomenon to medical use, treating rheumatics and paralytics with some success. Experiments that involved giving people electric shocks were quite common; on one occasion, 180 guards of the French king joined hands and static electricity was discharged along the line. Not surprisingly, all 180 jumped into the air at the same moment—which was doubtless amusing to the onlookers, if not to the guards.

Electrometers were used for measuring the quantity of electricity generated by early electrical machines. They came into use after the invention of the Leyden Jar in 1745–46. This was a simple device consisting of a metal-lined glass jar with a rod sticking into it through the cork. It could be used for storing the charge generated by an electrical machine; anybody who touched the central rod when the jar was fully charged would get a painful shock.

The electrometer shown here was made to a design attributed to John Cuthbertson. It dates from about 1799, a few years after Cuthbertson had returned to London from Holland, where he had worked as an instrument-maker since the late 1760s. This form of electrometer was used to regulate the discharge of a Leyden Jar. It consists of two vertical conductors, insulated from one another, each ending in a brass ball. The taller of the two, on the left, has a horizontal brass tube arranged to pivot from it; a circular ring, or weight, slides along this tube, which is bounded at each end with a brass ball. Above the tube is yet another ball, this one rigidly attached to the taller vertical conductor.

In use, the shorter conductor was attached to one side of the Leyden Jar and the longer conductor to the other. This created a difference of charge between the two stationary balls on the right, which were affixed to oppositely charged conductors. The pivoted ball was repelled from the ball above it (which had a like charge) and attracted to the one beneath it, which was oppositely charged. Whether the pivoted ball would descend far enough to make contact with the lower ball, thereby discharging the jar, depended on the position of the sliding weight; the nearer it was to the pivot, the greater had to be the attraction between the balls to overcome gravity and permit the tube carrying the ball to drop. By moving the slider along the pivoted tube, therefore, the quantity of electrical charge in the jar could be estimated and safely discharged.

The study of chemistry was transformed by Alessandro Volta's invention of the electrical battery in 1800. For the first time, a reliable soure of electrical power was available for experiments in electrolysis. In 1806 Sir Humphry Davy declared that many substances which had resisted chemical attempts to split them apart might be separated by electricity; two possible candidates were caustic soda (sodium hydroxide) and caustic potash (potassium hydroxide), both long suspected of harboring metals as part of their chemical structure. In 1807 Davy built an extra-strong battery, shown here, in an attempt to electrolyze these two substances.

The battery consists of a trough lined with pitch and divided into a series of cells by metallic plates. The dividers were made by joining together plates of zinc and copper face to face, so that each division is in fact a double plate, zinc on one side and copper on the other. Davy poured acid into the gaps between the plates, producing a series of cells which, connected together, made the most powerful battery yet constructed. On October 6, 1807, he passed the current from this battery through molten caustic potash and, to his joy, liberated small globules of metal; he had isolated potassium. A week later he isolated sodium from caustic soda, and the following year he went on to isolate magnesium, calcium, barium, and strontium.

In front of the battery trough are two small items that Davy used in some of his electrolysis experiments. One is a simple glass beaker with two platinum foil electrodes, and the other is a V-shaped electrolysis vessel, probably used in the electrolysis of water. This vessel has two electrodes, one in each arm of the V; when connected to a battery and filled with water, it would collect the oxygen given off in one arm and the hydrogen in the other.

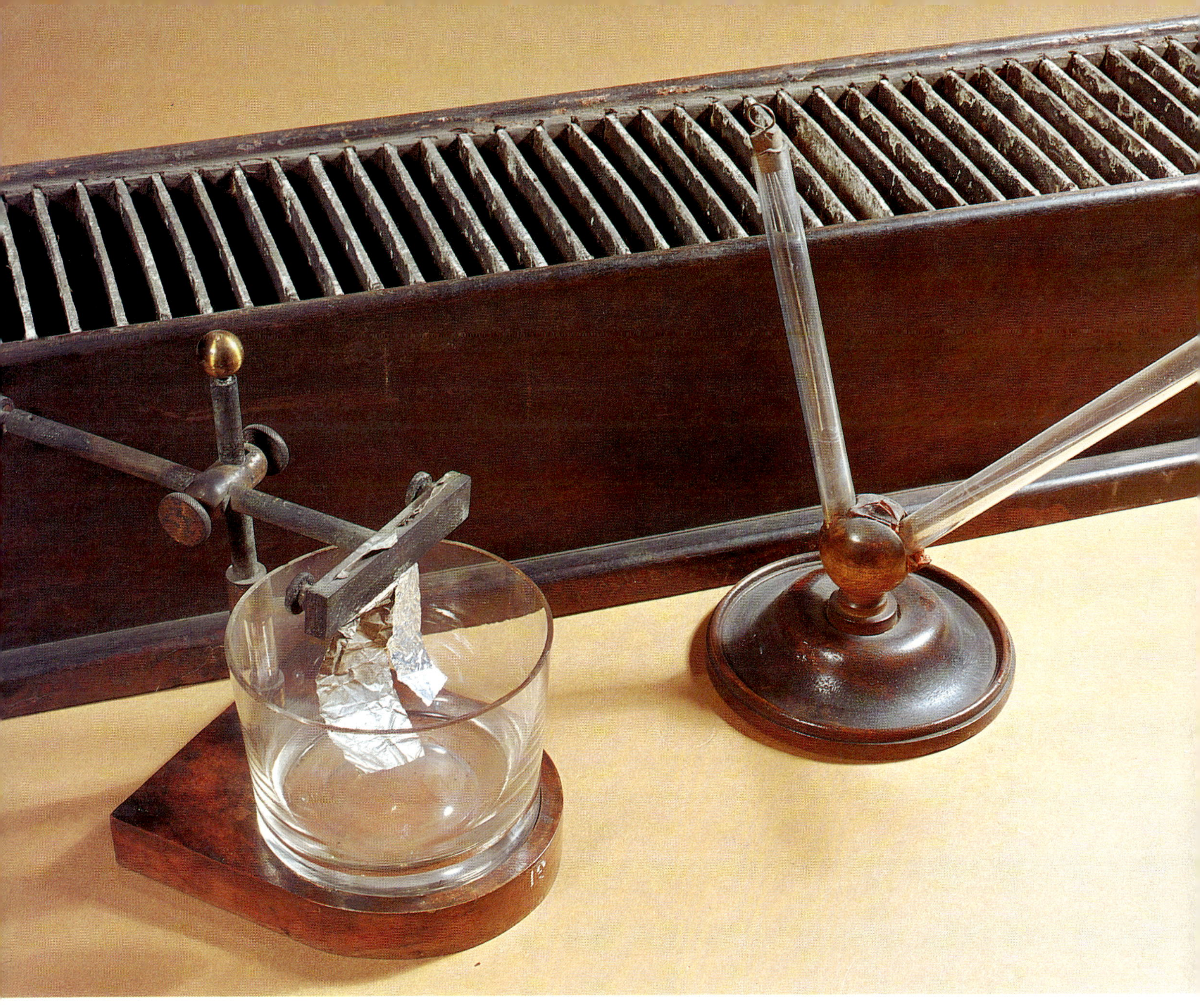

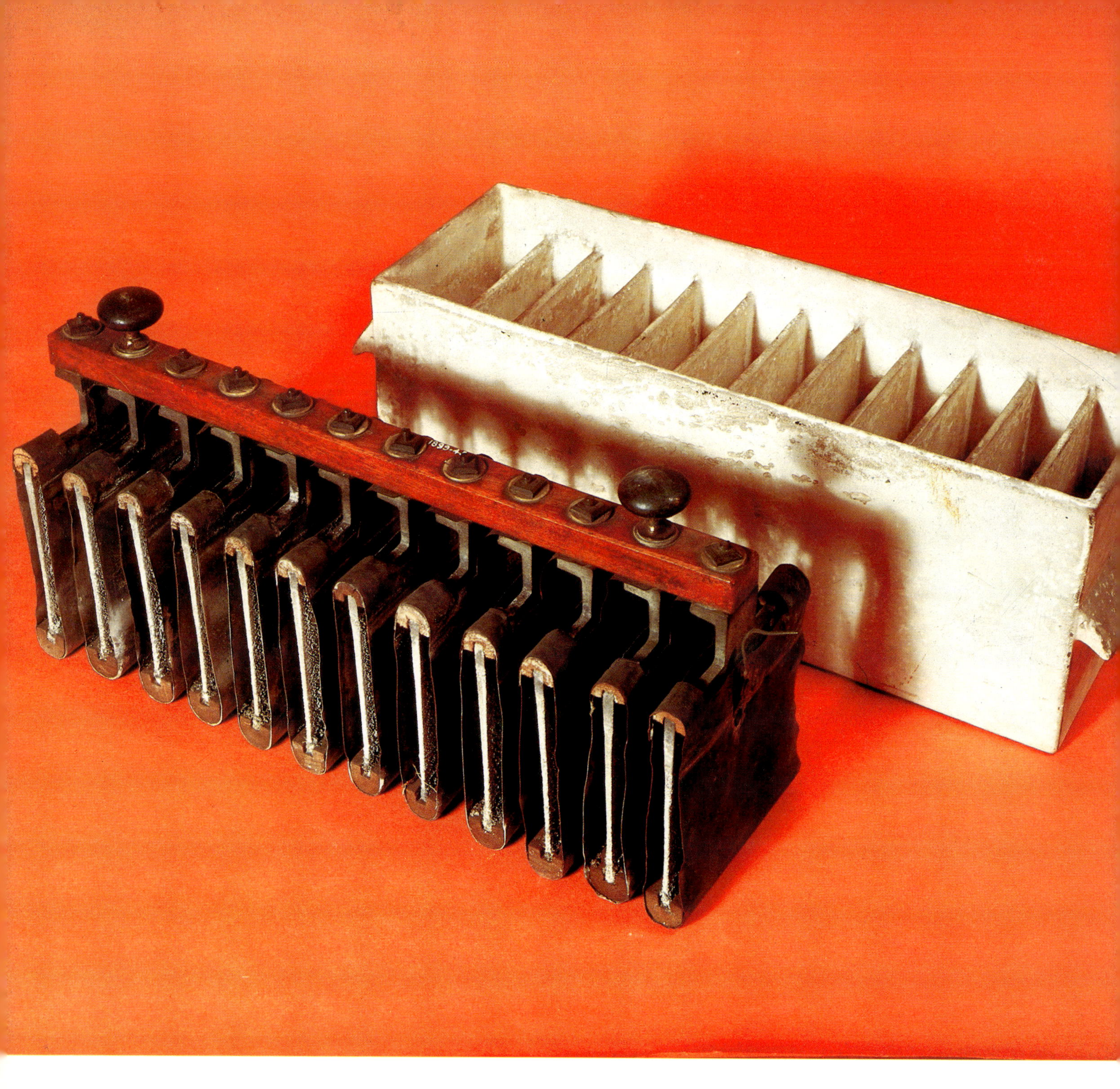

This battery was devised in 1815 by the English chemist and physicist William Hyde Wollaston, better known for his discovery of the rare metals palladium and rhodium. It is a voltaic pile, based on the discoveries made by Alessandro Volta fifteen years before. Volta found that an electrical battery could be made by mounting a series of copper-zinc junctions together in a salt solution; with this battery, which bears some recognizable similarities to those in use today, Wollaston greatly improved the convenience and simplicity of the voltaic series.

The case of Wollaston's battery consists of earthenware, divided into twelve separate cells filled with dilute sulphuric acid. Dipping into this acid are twelve copper-zinc pairs, mounted on a bar of varnished wood so that they can be removed easily from the battery case when not in use—as is the case here. In each individual cell the copper plate is made in the form of a U, so that it surrounds the zinc plate. Small pieces of wood support the zinc within the copper U without allowing it to make contact electrically. At the top of each of the copper loops is a tongue which makes contact with the zinc plate of the next cell along, thus creating a battery of twelve cells.

Wollaston was trained as a physician, practiced for seven years, and then decided to retire and devote himself to science. He discovered a method of making platinum malleable so that it could be used in laboratory instruments; this technique made him a fortune of £30,000, which he spent in further scientific pursuits. He was one of the first men to observe ultraviolet light, and he came close to discovering the laws of electromagnetic induction—but narrowly missed. It may have been a hint from Wollaston which put Michael Faraday on the right path to making that momentous discovery.

Quite independently and almost at the same time, Michael Faraday in London and Joseph Henry in Albany discovered that magnetism could produce electricity. This great discovery, electromagnetic induction, was to provide the foundation for a whole new industry. Faraday was a remarkable man; largely self-taught, he rose from humble origins to become perhaps the greatest experimental scientist of the nineteenth century. After finding fault with Sir Humphry Davy, he got a job at the Royal Institution in London, where his carefully recorded diaries and notebooks are still preserved.

On August 29, 1831, under the heading "Expts. on the Production of Electricity from Magnetism etc. etc.," Faraday recorded how he had taken an iron ring (shown on the far right) six inches in diameter and seven-eighths of an inch thick, and had wound on it two sets of coils, one on each half of the ring. One coil was connected to a galvanometer, the other to a battery. Whenever contact was made or broken in the second coil, a current momentarily flowed in the first. Thus had Faraday

invented the transformer, which is now universally used to convert low voltages to high and vice versa.

By October of the same year, Faraday had moved on to another vital experiment. He wound a long straight coil into a cylinder, so that a bar magnet could be slid inside it, and connected the end of the coil to a galvanometer. Whenever the bar magnet was moved back and forth, a current flowed. Faraday wrote in his notebook: "Hence here distinct conversion of magnetism into electricity." He had invented the first dynamo. The coil he used in this experiment is shown in the left foreground, with the iron bar magnet partially protruding. The other coils in the picture are of lesser historical significance; they were used by Faraday but when and for what purpose is no longer known.

Faraday called his discovery magneto-electric induction, though the term was subsequently reversed to electromagnetic induction. It is the principle on which all modern electrical machines are built; few discoveries have had a more far-reaching effect.

In electromagnetic induction, Faraday had discovered a new way of producing electricity; or was it, perhaps, a new phenomenon altogether? With this new source, there seemed to be five different ways of producing electricity: from batteries (voltaic electricity), from frictional electrical machines (static electricity), by heating the junction of two metals (the thermo-electric effect), from animals such as electric eels, and now from magnets as well. All of these different forms of electricity had features in common, but it was impossible to be certain that they were exactly the same phenomenon.

Faraday set out to prove that they were. He spent some time, using the apparatus on the right, demonstrating that he could produce a spark by electromagnetic induction. This was important because the production of sparks seemed to be characteristic of other forms of electricity.

The apparatus he used consists of a permanent magnet made of a series of bars of iron bolted together. On top of this lies another iron bar with handles on each end. Around this bar is wound a coil, one end of which terminates in a bare wire point and the other in a flat plate. To produce a spark, the wire is first placed in contact with the plate, and then the bar and coil, held by the handles, are jerked off the top of the permanent magnet. The effect is to change suddenly the magnetic flux and induce a current in the coil, simultaneously breaking the contact between wire and plate by the mechanical jerk. With luck, the circuit is broken just at the moment the induced voltage is at its peak, producing a spark between the end of the wire and the plate.

Up to the early 1830s, no accurate methods had been developed for measuring how much electricity any given source produced, and there was considerable uncertainty about whether different sources of electricity could produce the same effects. Using the apparatus shown here—technically a water voltameter—Faraday was able to measure the electrochemical effects of different sources of electricity.

The apparatus consists of a glass globe pierced by two electrodes, which terminate in a test tube supported vertically at the top of the globe. In use, the globe and tube would be filled with ordinary tap water. When the two terminals on either side of the globe are connected to a source of electricity, a current passes through the water, converting it into hydrogen and oxygen, which collect as gases in the top of the test tube. The amount of gas produced is shown by the movement of the water down the tube; assuming that the volume of water electrolyzed is proportional to the amount of electricity passed through it, it is possible to compare different sources of electricity.

In a later version of the same apparatus, shown in the background, Faraday provided two tubes with an electrode in each, so that hydrogen gathered in one and oxygen in the other. Using this type of apparatus, he was able to show that the products of electrolysis were produced in quantities proportional to their chemical combining weights. This observation became the basis of his second law of electrolysis. It was also at this time that Faraday, after consultation with the Reverend Wiliam Whewell, Master of Trinity College, Cambridge, coined a series of words which are still in universal use—electrode, anode, cathode, ion, anion, cation, electrolyte, electrolysis—to describe the processes he was studying.

1931-1122

Georg Simon Ohm was a German high school teacher who made several major contributions to the science of electricity, among which the discovery of how an electric current flows in a simple circuit was perhaps the most important. This is a reconstruction of the apparatus he used to make the discovery. The instrument consists of a thermocouple, a primitive galvanometer, and a pair of cups—the little eggcups on the left—containing mercury, into which the wire under investigation could be dipped. The thermocouple, which provides the power for the device, consists of a rectangular strip of bismuth, joined at the ends to two strips of copper. When bismuth-copper junctions of this sort are heated to different temperatures, an electromotive force is created; so Ohm dipped one end into ice and the other into steam, in the tripod-supported vessels on the right.

The galvanometer consists of a needle supported by a fine gold wire, mounted above the copper strip. The needle was magnetized, so it normally pointed north. When the circuit was completed by dipping the two ends of the piece of wire (not shown here) into the eggcups, the copper strip carried a current that varied according to the resistance of the wire. This current created a magnetic field, which deflected the magnetic needle. The needle, mounted on a torsion wire, could be restored to its original position—pointing north—by twisting the brass cap at the top, to which the torsion wire was attached. The greater the current flowing, the more the cap had to be turned to restore the needle to its original position; thus the amount by which it had to be turned was a measure of the current.

Using this device, Ohm studied the relationship between electromotive force, resistance, and current, and formulated in 1826 his classic equation, $V = iR$: Ohm's Law. The secret of his success was to use the thermocouple to produce electromotive force, instead of the unreliable electrical batteries of the day which had been used by other researchers.

1876-45

James Prescott Joule was the grandson of a wealthy Salford brewer, and his name is immortalized in the term used to measure units of work. Joule's major discovery, to which he devoted many years of study, was that work is converted to heat according to a fixed ratio known as the mechanical equivalent of heat. He is notorious for having taken along a thermometer on his honeymoon in order to measure the difference in the temperature of the water at the top and bottom of a waterfall; since the falling water loses its energy of motion when it stops at the bottom, Joule reasoned that it ought to be converted to heat. (His thermometer, unfortunately, was too insensitive to detect the very small change in temperature.)

Patient and methodical, Joule carried out many experiments in electromagnetism, using a powerful magnet that he designed himself. It consisted of a piece of iron one inch thick curved around into the shape of a horseshoe and wound with sixty-eight yards of copper wire. This electromagnet, which weighed over one hundred pounds, was placed in a wooden box (thoughtfully supplied with carrying handles) and held in place by two straps of brass across the top. Two brass terminals at the front provided the means of connecting the coils to batteries.

In a series of experiments reported in *Philosophical Magazine* in 1852, Joule showed that the power of the magnet was proportional to the current flowing in its coils. Using a pair of tapered poles (shown in front of the magnet), he demonstrated that bismuth was diamagnetic—that is, it tended to be repelled by the magnetic field—and that its diamagnetism was not inherent in the bismuth but induced by the electromagnet. Joule's most important discovery in magnetism was magnetostriction—the property of an iron bar to change its length when magnetized. Although it was of no apparent importance when it was discovered, the principle of magnetostriction is now applied to the production of ultrasonic sound waves.

For the first half of the nineteenth century, units of electrical resistance were defined in an unsatisfactory and rather arbitrary way: by reference to standard conductors in the form of lengths of wire. In order to put the matter on a sounder footing, it was necessary to devise an apparatus that could measure electrical resistance in terms of the absolute units of length and time. This coil, designed in 1863 by Professor William Thomson (later Lord Kelvin) for the Committee on Standards of Electrical Resistance, was designed to do just that.

The apparatus consisted of a copper coil, held vertically on bearings so that it could be rotated at a constant speed in the earth's magnetic field. At the center of the coil was suspended a magnetic needle; when the coil was rotated, the induced magnetic fields caused the needle to deflect. From the dimensions of the coil, its speed of rotation, and the degree of deflection of the magnet, the coil's electrical resistance could be determined in absolute measure—that is, in terms of length and time. This research was carried out on behalf of the British Association for the Advancement of Science, and the unit of resistance so defined was called the B.A. unit. It could then be used to calibrate the standard resistors.

The rotating coil was first used in 1863-64 by J. Clerk Maxwell and Balfour Stewart at King's College, and later by Lord Rayleigh at the Cavendish Laboratory in Cambridge. In fact, subsequent, more accurate determinations of electrical resistance have shown that the B.A. unit of 1863-64 was not, as it should have been, exactly 1 Ohm. Inaccuracies in the experimental method had introduced small errors, so that the B.A. unit was in fact only 0.9866 Ohms.

Nikola Tesla, a Hungarian who emigrated to the United States in 1884, was the man who made alternating current a practical proposition for the distribution and use of electricity. In doing so he clashed with Thomas Edison, the most powerful and influential inventor of the day, who favored direct current. (In an alternating current system, the direction in which the current is flowing reverses itself many times a second; in direct current it flows in the same direction all the time.)

The major advantage of alternating current is the ease with which it can be transported over long distances by the use of transformers. The loss when electricity flows through wires is proportional to the square of the amount of current flowing; so for long-distance transmission, it is best to keep the current as low as possible. This can be done with transformers, which increase the voltage of an electricity supply while reducing the current. Today, long-range transmission of electricity takes place at enormously high voltages—up to 750,000 volts. At its destination, the electricity is passed through more transformers to bring its voltage down to the domestic level of 250 or 110 volts.

The Tesla coil shown here has been set up to illustrate the principle of the transformer. The primary winding, on the clear plastic sleeve, consists of just a few coils of copper wire. The secondary winding, the main coil, consists of many turns. When a voltage is applied to the primary windings from the induction coil through a series of condensers, it produces an electrical oscillation in the primary coil, to which the secondary winding responds. Because there are many more windings on the secondary coil, the voltage which it picks up from the field produced by the primary is greatly magnified. This kind of transformer also demonstrates the principle of electrical resonance: So high is the voltage in the secondary that it creates an electric field strong enough to cause a glow in the rarefied gases of the light bulb hanging above the coil.

The first electric telegraphs used a receiver which consisted of a needle deflected by the current to different positions on a dial. The operator watched the needle, calling out to an assistant each word as it was spelled out—a procedure that was both expensive and subject to frequent mistakes. Charles Bright, a telegraph engineer, improved the system by making use of the sense of hearing instead of sight; he called his invention the acoustic telegraph.

The idea was simple enough: Instead of a needle deflecting to right or left, Bright arranged for the incoming signals to operate one or the other of two bells, each with a different note. Thus the operator could simply listen, his eyes on his notepad, and transcribe each letter as it was transmitted to him. This procedure was more reliable and accurate than the visual system, and could be handled without an assistant. The apparatus illustrated here shows Bright's double-bell receiver, patented in 1855.

Bright's greatest triumph, and his principal claim to posterity, was the laying of the first telegraph cable across the Atlantic. He became chief engineer to the Atlantic Cable Company in 1856 and designed the complex machinery needed for paying out the cable from the ship. The first attempt, made in 1857, failed after 380 miles of cable had been laid, when a break occurred in 2,000 fathoms of water. The second attempt, in 1858, also failed, but a third expedition was quickly mounted and in August 1858 success was finally attained. Charles Bright was knighted by the Lord Lieutenant of Ireland when he returned home after this great achievement. His triumph, however, was short-lived. The cable's insulation rapidly failed, and in October it stopped working altogether, though not before Queen Victoria had been able to send a message to the President of the United States. It was to be another eight years before a reliable transatlantic cable could be laid, using Brunel's ship, the *Great Eastern.*

RIGHT'S PATENT.
HENLEY
203
P.O.
M Nº 1950.

Professor Charles Wheatstone of King's College, London, was a pioneer of the electrical telegraph; together with William Cooke, he was responsible for its introduction into Britain. Their first public telegraph service was installed along the Great Western Railway, and by 1842 it extended from Paddington to Slough.

In 1845 the system became famous when it led to the arrest of a murderer. A woman had been killed in Slough, and the man believed to be responsible was seen getting aboard a Paddington-bound train. A full description was telegraphed ahead of the train; the man was arrested, charged, and subsequently convicted and hanged. (A remarkable parallel took place fifty-five years later when the murderer Dr. Crippen was arrested in the United States as he alighted from a transatlantic liner. In that case, descriptive information had been transmitted by wireless telegraphy to the New York police.)

Shown here is one of Wheatstone's major contributions to the history of the telegraph: the punched-tape perforator, transmitter, and pen-tape receiver, first patented in 1858. A good telegraph operator, working in the code invented by Samuel F. B. Morse, could send out no more than twenty-five words a minute along the telegraph wire, which was too slow a rate to get a good return on the heavy investment of laying cables. The Wheatstone system enabled messages to be prepared in advance in the form of a punched tape (an idea first proposed by Alexander Bain in 1847), and then transmitted automatically at three times the speed a human operator could manage. At the other end, Wheatstone's receiver translated the electrical dots and dashes into marks on a strip of paper. This elegant and simple apparatus transformed the efficiency of the telegraph.

SIEMENS BROS & Co
LONDON

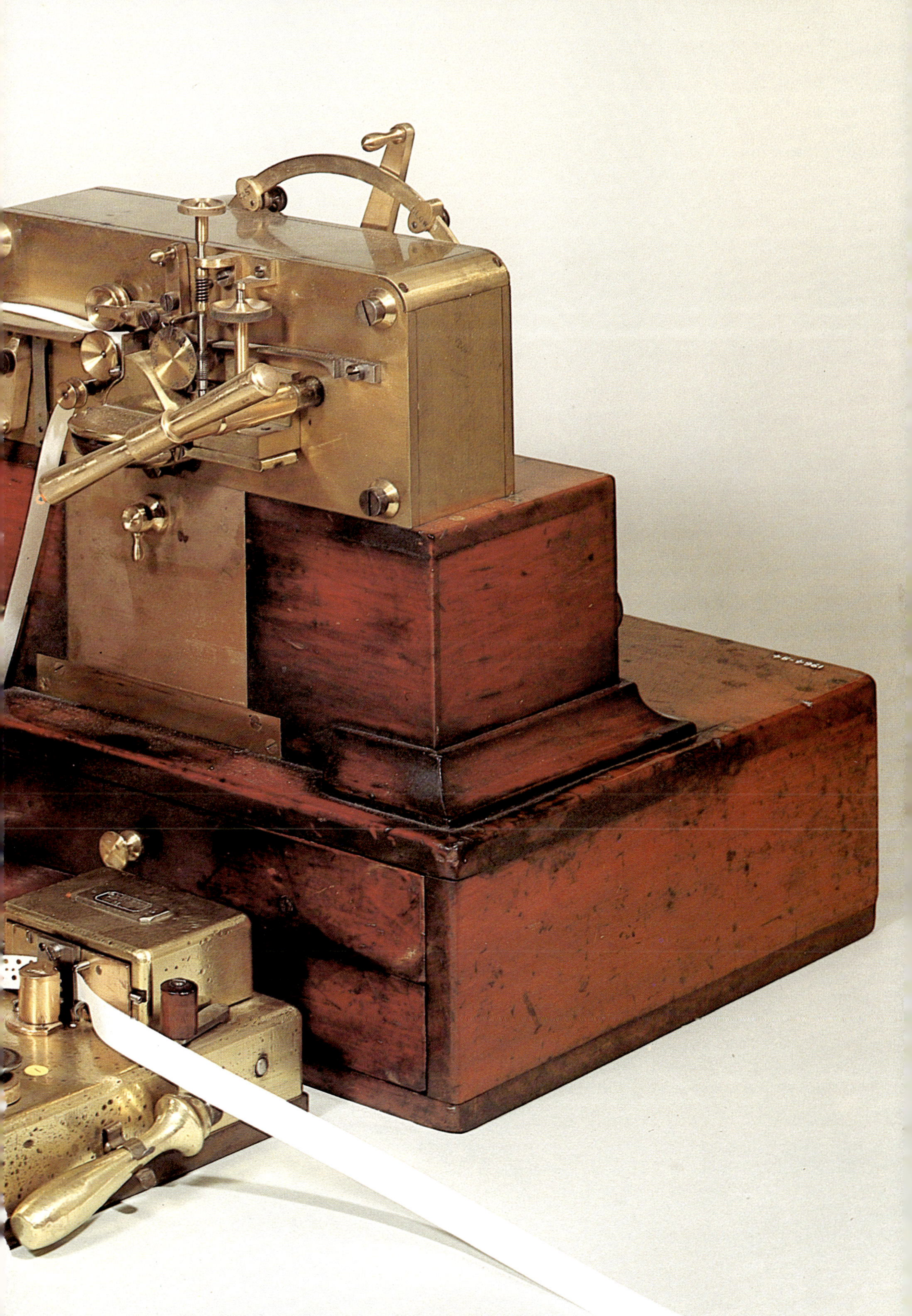

LIFT

After the telephone had been patented by Alexander Graham Bell in 1876, the need quickly arose for a switchboard to connect one telephone subscriber to another. One of the earliest of these switchboards, designed for fifty subscribers and made by Jones Bros. of Cincinnati, Ohio, in 1879, is shown opposite. It was one of the first switchboards to be installed in Norway by the Bell Telephone Company, and was in use in Drammen between 1880 and 1889. It had previously been used in the United States.

In this switchboard, each subscriber's line terminates in one of the indicators, which are arranged in four rows with wires feeding into them. The red "indicator drops" are normally held in the upright position by an electromagnet. When a subscriber calls, the armature of the electromagnet is withdrawn, allowing the drops to fall and signal the operator as to which subscriber is calling. Below each indicator is a plug socket; to connect two subscribers, all that is necessary is to plug a double-ended cord into the appropriate sockets.

Above the indicators is an additional panel of transfer sockets. A large exchange might have many such switchboards, each attended by a single operator. If a subscriber coming in on one switchboard was trying to ring one whose line ended in another switchboard, the transfer sockets could be used to connect the two switchboards together. This system worked well until the number of subscribers reached a certain critical level—about 500. After that, the task of connecting one switchboard with another became almost impossible. Further technical developments were needed before really large exchanges could be coordinated.

Professor David Hughes, born in London of Welsh parents in 1831, was the inventor of the microphone, and very nearly invented the radio. Taken by his parents to the United States at the age of seven, Hughes eventually became Professor of Music at a small college in Bardstown, Kentucky. He was interested in invention and devised a number of improvements on the electric telegraph, including a printing machine, which he patented in 1856. Hughes returned to Europe and saw the installation of his telegraph printing apparatus, first in France in 1861, and by 1869 throughout Europe. He resettled in London in 1877 and lived there until his death in 1900.

In 1878, Hughes demonstrated to the Royal Society the first microphones—a term he invented by analogy to the microscope, for his devices magnified sounds just as the microscope magnified objects. He had discovered that electrical conductors in intermittent contact will pick up and reproduce sounds. The simplest such microphone consists of no more than three nails, two lying parallel and the third lying across the other two and loosely connecting them. Other versions of the invention shown here had springs to hold the conductors together under pressure, or consisted of tubes of glass, quill, or bamboo, filled with granules of zinc or silver. These first microphones were discovered among Hughes's effects after his death in 1900 and presented to the Science Museum in London by his widow.

Hughes believed, correctly, that the amplifying effect was due to differences of pressure, caused by sound waves, at the points of contact of the conductors. The best results were achieved when there were many points of contact between the conductors, such as might be achieved with finely granulated carbon.

Immediately after Hughes made public his invention of the microphone, Thomas Edison, the American inventor, launched an accusation of piracy against him, claiming that Hughes's microphones were simply copies of Edison's "transmitters." So heated did the disagreement become that William Thomson (the eminent Scottish physicist who later became Lord Kelvin) felt obliged to intervene on Hughes's behalf, declaring that Edison had let himself "be hurried into an injustice." At this distance, it is hard to judge who was right; but it does seem that Hughes had a greater theoretical understanding of the microphone than Edison did, as well as the ability to explain correctly how it worked.

Hughes's microphones took various forms. The simplest were made with three nails; a more complex version, illustrated here, consisted of two weighted carbon rods suspended by paper strips so that they lay horizontally in contact with two vertical carbon rods.

In 1879, the year after Hughes had demonstrated his microphones to the Royal Society, he unwittingly discovered radio waves. He noticed that his microphones responded whenever the current flowing in a nearby coil was interrupted, even though there was no electrical connection between them. He demonstrated the phenomenon for visiting scientists by walking up and down Great Portland Street in London with a telephone receiver in his hand and showing that he could hear sounds as distant from the source as 500 yards. In February 1880 he showed the effect to the President of the Royal Society, William Spottiswoode, who was accompanied by T.H. Huxley and Sir George Stokes, both very eminent scientists. They watched for three hours, at the end of which Stokes declared that all the results could be explained by well-known principles of induction.

Hughes, somewhat discouraged by this verdict, did not persevere with his experiments. If he had, he would almost certainly now be regarded as the discoverer of radio waves.

Count Rumford, born Benjamin Thompson in Woburn, Massachusetts in 1753, was a pioneer in the study of heat and an enthusiast for the application of science toward human betterment. He backed the wrong side in the American Revolutionary War, and after the colonists had won he was forced into permanent exile in Europe. In 1783 he went to work for the Elector of Bavaria, where he studied the heat generated when brass cannons were bored out during manufacture. He concluded, correctly, that the mechanical motion of the boring tool was being converted to heat. This was not what most of his contemporaries believed, and it was many years before Rumford's view was accepted. In 1790 the Elector made him a count and he took the name of Rumford, a town in Massachusetts near which he had an estate.

Rumford is best known today for having founded the Royal Institution of Great Britain, the first publicly supported scientific laboratory in the world. It was in 1796 that he put forward the idea for an institution "for introducing and bringing forward into general use new inventions and improvements, particularly such as relate to the management of heat and the saving of fuel, and to various other mechanical contrivances by which domestic comfort and economy may be promoted." (As it happened, the Royal Institution turned out rather differently, as a place where fundamental new discoveries were made.) Rumford made these simple burners, one with several wicks (the polyflame burner), probably for heating purposes in the laboratory rather than for lighting.

Rumford traveled easily between Britain and France, despite the Napoleonic Wars, and in 1804 he met and married the widow of the French chemist Antoine Lavoisier. Rumford had also been married before, but his first wife, whom he had left behind in America, had died. Rumford and Madame Lavoisier were not happy together and separated four years later. Rumford ruminated, rather unkindly, that Lavoisier had perhaps been lucky to have been separated from his wife by the intervention of the guillotine.

In 1792 Count Rumford wrote to Sir Joseph Banks, President of the Royal Society, describing "A method of Measuring the Comparative Intensities of the Light emitted by Luminous Bodies." This was his shadow photometer, shown here, an instrument which enabled the candlepower of any two sources of light to be compared.

The instrument consists of two V-shaped channels, through which light from the two sources is directed onto a screen. Between the lights and the screen are two circular pegs with vanes attached to them; by rotating these pegs, shadows of the two lights are cast onto the screen in such a way that they lie directly alongside each other. When the intensity of the two shadows matches, then the sources of illumination are of equal brightness. In the photograph, to illustrate the principle, the photometer has been illuminated on one side with green light and on the other with purple, casting onto the center of the screen one purple shadow and one which is dark green, almost black.

In practice, the photometer would not have been used with lights of different colors but with two sources of white light, one a standard source and the second a source whose intensity was to be calculated. The two lamps would be mounted on arms which could be moved in and out by means of handles (not shown here) controlled by the operator. When the sources were moved back and forth until the two shadows matched in intensity, the candlepower of the unknown source could be calculated by means of the inverse square law. This states that the power of a lamp varies in proportion to the inverse square of its distance from the screen. Thus if the shadow from the unknown lamp matched that from the standard source when it was twice as far away from the screen, then the unknown lamp would be four times as powerful as the standard source.

Sir Humphry Davy is believed to have invented the carbon arc lamp in 1801—although, curiously, it is not an invention he himself ever claimed. However, he was the first to describe experiments using the lamp, in which an electric arc is struck between two carbon electrodes. Shown opposite is an arc of Davy's time, with two carbon elements held in supports so that they can be touched together momentarily and then separated to start the arc. The elements themselves are not contemporary with the rest of the apparatus; they are more modern carbon electrodes, whereas in Davy's day they would have been made of charcoal.

This fragment of apparatus is believed to have been used by Davy to disprove the belief, widely held in his day, that heat cannot travel through a vacuum. Those who subscribed to this view acknowledged that heat reaches the earth from the sun through a vacuum, but argued that once it was within the earth's influence it needed the atmosphere to transport it. Davy disproved the hypothesis through the following experiment: He took a bell jar and mounted in its neck a source of heat (a carbon arc is shown here, though in the original experiment a platinum element was used). A small piece of the neck of the bell jar can still be seen, although the rest is gone. Inside the bell jar Davy mounted a thermometer or thermoscope, whose blackened bulb is visible. Hc then evacuated the bell jar, connected up the heat source, and showed that the thermometer registered an increase in temperature, even though there was no air to transport the heat from the heat source to the bulb of the thermometer.

Davy is known to have carried out this experiment with a platinum element, as he recorded the fact in his notebook. He may very well have used the carbon arc as a variation of the experiment, but he never made specific mention of it.

As the use of coal expanded in the eighteenth century, mining became a major industry, forcing miners to dig deeper and deeper in search of rich veins of coal. Ventilation at these depths was inadequate, and frequent underground explosions were caused when the flame of a candle or an oil lamp ignited pockets of methane gas, commonly known as "fire-damp." By 1812 the explosions had become so common that an advisory body—the Sunderland Committee—was set up in the Northeast of England to try to do something about it.

The committee enlisted the aid of Sir Humphry Davy, a professor at the Royal Institution. By the end of 1815 he had discovered the laboratory conditions under which methane gas explodes, how inflammable it is, and how it behaves when mixed with air in various proportions. He studied the passage of flames through narrow apertures, and found that metal gauze made an effective barrier, cooling the flame to a temperature below that which will ignite methane.

The first site tests of the new miner's lamps that Davy designed were made in January 1816 in Hebburn Colliery, near Newcastle. John Buddle, a local mining engineer, reported: "To my astonishment and delight, it is impossible for me to express my feelings when I first suspended the lamp in the mine and saw it red hot. If it had been a monster destroyed I could not have felt more exultation than I did." In fact, Buddle was exaggerating. Although legend has it that Davy's lamp banished the risk of explosions in the mines, in reality it did no such thing. Coal owners drove even deeper shafts and reaped bigger profits while miners continued to die in explosions. In the twenty years after the first successful tests at Hebburn, deaths down the mines increased by twenty percent in England.

Illustrated here are three early commercial lamps, manufactured soon after the 1816 tests. They have a double layer of wire gauze at the top, because it was found that a single layer was sometimes not enough to keep the flame cool. The middle lamp also has a "bull's-eye" lens to focus the rather dim light which these early lamps produced.

47

The pole lathe was a medieval invention which survived almost unchanged for five hundred years. It used a cord, wound around the piece to be worked, to turn it against the cutting tool. One end of the cord was connected to a foot treadle, while the other end was attached to a springy pole bent over in a bowlike arch (not visible here). On the downstroke of the treadle, the piece rotated towards the craftsman and he applied his tool to make a cut. At the bottom of the stroke, he took his weight off the treadle and the pole sprang back to its starting point, ready for another stroke.

The pole lathe was only really suitable for wood turning, and then only for pieces of a modest size. The tool was not fixed, but held in the hand or fitted with a handle long enough to bear against the workman's shoulder. The first illustrations of pole lathes appear in the thirteenth century; they are little different from the lathe shown opposite, which dates from about 1800. The lathe bed consists of two oak beams mounted on posts, and the headstocks are solid blocks of oak fitted with pointed iron "centers," between which the workpiece rotates. The tool, which rests on a crude piece of wood, can cut only when the workpiece is rotating toward the operator.

The survival of such a simple tool for so long was a remarkable example of how advanced technology can simply pass the rural craftsman by. L. T. C. Rolt, an English historian of technology, has recorded how just before the Second World War he watched George Lailey, the last of the Berkshire bowl-turners, making elm bowls in a small shed on Bucklebury Common. The pole lathe he used hardly differed at all from its medieval ancestors.

The founder of the modern machine tool industry was Henry Maudslay, born at Woolwich on the banks of the Thames in 1771. By the age of twelve he was at work at Woolwich Arsenal, where he soon showed an astonishing aptitude as a metalworker. He moved to the workshop of Joseph Bramah, where he helped to solve the problems involved in making a revolutionary new lock. In 1797 Maudslay rashly asked for a raise, was turned down, and promptly left to set up his own workshop, from which there soon began to flow a series of superb machines, ancestors of today's machine tools.

By 1810, when he designed the lathe shown on the opposite page, Maudslay had founded the firm of Maudslay Sons and Field in an abandoned riding school on the outskirts of London. This tool is one of the very few he ever offered for sale; its price was £200. The lathe slides along a triangular bed, and is driven by a treadle attached to a pulley (far left). The belt drive on the pulley has five separate positions, giving the machine five speeds. The bronze gear wheels between the pulley and the headstock of the lathe are used for screw cutting. They drive a shaft running through the center of the bed of the machine, which advances the tool at a fixed speed.

Maudslay's standards of accuracy in the cutting of metal were revolutionary. He produced a brass screw seven feet long with an error of only one-sixteenth of an inch over its full length, and designed a micrometer capable of measuring deviations of one ten-thousandth of an inch. Almost single-handedly, he turned metalwork from a craft into an industrial science.

LIFT

The first machine tools capable of accurate and delicate work were ornamental lathes, often known as rose engine lathes. They were used for cutting wood, ivory, or soft metals into attractive patterns, or for making copies at varying scales of coins or medallions. They first appeared in the sixteenth century—the earliest known illustration appeared in 1569—and by the mid-eighteenth century, the date of this German lathe, they had reached a state of remarkable refinement.

In the rose engine lathe, cams and templates control the movement of the workpiece and tools, producing intricate and delicate patterns. It became something of a fashionable hobby among the wealthy to own and use an ornamental lathe; Louis XVI of France had one. As a result of this aristocratic patronage, the lathes themselves became objects of beauty, decorated with elaborate scrollwork and rococo brackets, very different from the down-to-earth machine tools that began to be produced later in the century. Rose engine lathes were not used by ordinary craftsmen, who managed to produce work of perfectly acceptable quality on much simpler machines such as the pole lathe. A wood-turner could never in a lifetime have earned enough to buy a rose engine lathe; and besides, the quality he could have achieved with it would have been far greater than the market demanded. Only when the Industrial Revolution was well underway did greater accuracy become absolutely necessary, and then a new type of machine tool, much stronger and more scientifically designed, emerged.

The cutting tool of an ornamental lathe was held against the workpiece by cams working against weights or springs—a satisfactory arrangement for cutting soft materials, but unsuitable for most metals. Industrial machine tools would need rigidly fixed cutting edges to stand up to the rigors of heavy use.

The English physicist Sir William Crookes invented his radiometer in 1874. It is an ingenious little toy, of no huge importance to the history of science, but instructive and attractive. Small models very similar to this one can still be bought in toy shops today.

The radiometer consists of an evacuated glass globe with a pivoted "paddle wheel" arranged to rotate about a vertical axis inside it. The circular paddles are blackened on one side, and polished or silvered on the other. When the radiometer is exposed to sunlight, or any other form of radiation, the paddle wheel rotates; the brighter the sun, the more rapidly it spins. The direction of the spin is such that the blackened surfaces of the paddles appear to recede from the source of radiation, as if the radiation itself was pushing the paddle wheel around. To Crookes that did, indeed, seem to be the explanation.

Despite appearances, however, that is not how a radiometer works. If a perfect vacuum could be created inside the globe, the rotation of the wheel would stop, even though the radiation would be just as strong. What actually happens is that the residual air molecules left in the globe after evacuation provide the mechanism for driving the wheel. Radiation falling onto the white or silvered surface of the paddles is reflected, while that falling on the dark side is absorbed. This creates a difference of temperature between the two sides; the air on the black side is heated, causing its molecules to move about more rapidly. Because the molecules on the black side are moving more quickly than those on the white, they exert a greater force when they collide with the paddle, thereby driving it around. So although at first sight the radiometer appears to be suggesting that light consists of tiny particles which drive the wheel around, what it really demonstrates so elegantly is the truth of the molecular theory of gases.

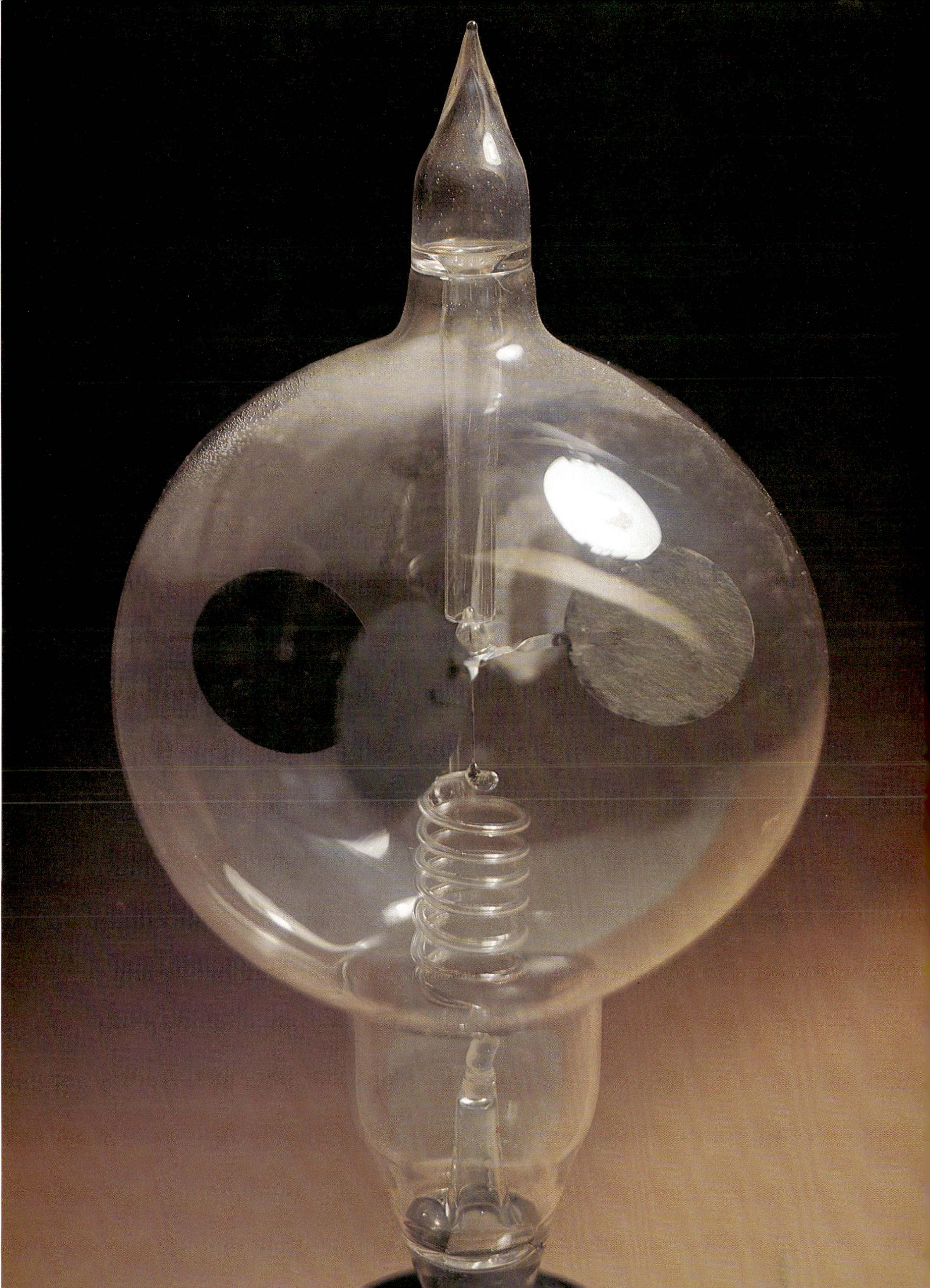

Sir William Crookes designed this string-and-sealing-wax arrangement in 1879 to study the behavior of cathode rays. It was known that when a negative electrode—a cathode—was placed in an evacuated glass tube under strong electric voltage, radiation and luminescence appeared around it. In 1876 the German physicist Eugen Goldstein had called such radiation cathode rays.

Crookes's apparatus, known as a Crookes tube, is designed to investigate the behavior of the rays at various gas pressures, and to subject them to a magnetic field. The cathode is at the bottom of the tube; directly above it is a mica screen with a central slit, through which a narrow beam of cathode rays can pass. The rays bounce off the glass walls of the tube and fall on the fan-shaped fluorescent screen, which is inclined at an angle to their path so that the rays produce a luminous line—by exactly the same principle, incidentally, as a picture is produced on a television screen. At the bottom of the stem of the tube is some caustic potash, which gives off moisture when heated and reabsorbs it as it cools. By controlling the temperature of the tube, Crookes was able to increase or decrease the amount of the water vapor, and hence the vacuum, at will.

Crookes found that when a bar magnet was brought near the tube, the beam of cathode rays was directed into a spiral, and that a horseshoe magnet bent the path into a curve. The direction of the deflection was that which would be expected if the cathode rays consisted of a stream of negatively charged particles. Crookes also observed that the curved path produced by a magnetic field grew flatter as a better vacuum was achieved, demonstrating that the initial velocity of the particles increases as the vacuum becomes purer.

The Dewar vacuum flask, today known universally by the brand name of Thermos, was originally invented not to keep hot drinks hot, but to keep liquefied gases cold. Its inventor was Sir James Dewar, a professor at the Royal Institution, who was engaged in research on liquid oxygen and nitrogen during the 1890s. These two liquids can only be produced at very low temperatures, and Dewar found it difficult to store the fluids once they had been produced, because they quickly took up heat from their surroundings and vaporized.

Dewar started simply enough with a double-walled vessel (top left), which improved the insulation of the inner vessel in much the same way as double window glazing. His next step was to evacuate the space between the two vessels (top right), which eliminated heat loss from convection. Then, to minimize radiation losses, he covered the outside of the inner vessel with silvering, so that it reflected heat back into the vessel rather than absorbing it.

The vessel in the foreground is divided into three sections, with a common space surrounding the inner chambers. One of the three sectors is plain glass; one has its evacuated space filled with crumpled gold foil; and the third has flat gold foil on the inner and outer surfaces. Using this vessel, Dewar could compare the insulating properties of different vacuums and different arrangements of the foil. By a process of experimentation he developed the Dewar flask—a double-silvered, evacuated vessel which was supremely well insulated and which has survived in unchanged form to this day.

Dewar never patented his flask; had he done so, he would have become extremely rich. And curiously, although Dewar is universally acknowledged as the inventor of the vacuum flask, the principles involved had actually been suggested some forty years before by L. Gompertz, in a book called *Mechanical Inventions and Suggestions on Land and Water Locomotion.* Gompertz, who published his book in 1850, never put the idea into practice, and even expressed a doubt that it would work.

Before the advent of the friction match in 1827, the wealthy might have owned a table lighter like this one to light their lamps and their cigars. This Temple of Vesta, named after the Roman goddess of the hearth and household, was made in 1807. Vesta can be seen at the open door of her temple; the lion, which normally sits immediately in front of the door, has been removed and placed at the base of the temple in order to show how the mechanism works.

Underneath the elaborate decoration, the Temple of Vesta is a Kipp's apparatus for the production of hydrogen. A cylinder of zinc, suspended in a central glass vessel, makes contact with dilute hydrochloric acid contained in a surrounding vessel, liberating hydrogen in the reaction. The two vessels are arranged so that when the volume of hydrogen so produced increases, the pressure of the gas drives the acid out of the vessel containing the zinc, controlling the reaction and inhibiting the production of hydrogen. In this way hydrogen is liberated in a safe and regulated manner. The base of the temple contains an electrophorus, a source of static electricity. It consists of a disc of resin which is charged by rubbing it occasionally with fur.

When the little knob on the right of the base is pressed, two things happen. First, a gas tap is opened, allowing hydrogen to escape through the nozzle visible in front of the goddess Vesta. Second, the charged resin disc is brought into contact with a wire, which is carried in an insulated sheath through the base board and ends close to the gas nozzle. This produces a spark which ignites the hydrogen, creating a flame. In use the lion would be placed over the nozzle and wire so that the flame would issue forth from its mouth.

Devices like the Temple of Vesta were not uncommon, though seldom were they decorated so handsomely as this specimen. They became obsolete after the first friction matches were sold by John Walker, a pharmacist from Stockton-on-Tees in northeastern England, in 1827.

This charming lamp, really more a scientific toy than a source of light, illustrates an important discovery made by the German chemist Johann Dobereiner in 1823. Observing that when a jet of hydrogen was directed at a stream of pulverized platinum, it spontaneously burst into flame, he calculated that the platinum was acting as a catalyst in the chemical reaction between the hydrogen and the oxygen in the air. Normally a spark or a flame is needed to make hydrogen and oxygen react, but in the presence of platinum—which facilitates the reaction without being changed itself—combustion takes place spontaneously.

Dobereiner, who had been appointed professor of chemistry and pharmacy at Jena in 1810, used the discovery to make a self-igniting lamp. Its glass vessel contains zinc and sulfuric acid; when the tap is opened, the acid comes into contact with the metal and reacts with it, producing hydrogen. The gas is released through the urn on the top of the lamp, which contains the platinum. The flame which results from the catalytic reaction, described above, ignites an oil lamp contained in the lid of the instrument. The lamp had some drawbacks. The platinum was expensive, and was soon rendered ineffective by impurities in the hydrogen. Nevertheless, by 1828 some 10,000 Dobereiner's lamps were in use in England and Germany. Unfortunately, the inventor neglected to patent the design; as he had once remarked, he "loved science more than money."

The principle on which the lamp worked turned out to be a very important one. By encouraging reactions to proceed at lower temperatures and pressures—and by making possible some chemical reactions which would not work at all without it—platinum has played a key role in the development of the chemical industry, and it remains an essential element of many processes today.

The stereoscope was a scientific toy which captured the imagination of Victorian society by producing extraordinarily convincing three-dimensional images. In 1832, Sir Charles Wheatstone designed the first instrument to give the effect of stereoscopic relief. Sir David Brewster, the Scottish scientist (best known to the public for his invention of the kaleidoscope in 1817), designed this form of "lenticular" stereoscope in 1849. Neither instrument attracted any interest outside scientific circles. Fortunately, Jules Duboscq of Paris took up the idea and exhibited a stereoscope at the Great Exhibition of 1851, held in Hyde Park, London. Queen Victoria herself took a fancy to it, and as Brewster himself wrote in 1856: "The demand became so great that opticians of all kinds devoted themselves to its manufacture, and photographers found it a most lucrative branch of their profession, to take binocular portraits of views to be thrown into relief by the stereoscope."

The stereoscope achieves its illusionary effect through the superimposition of two images of the same object, taken from slightly different points of view. The two images mimic the action of the eyes—which also focus on an object from separate perspectives—to produce a remarkably effective three-dimensional imagc. In 1849 Brewster proposed a twin-lens camera capable of taking both images simultaneously; the first one was made by John Benjamin Dancer of Manchester in 1853. It quickly came into common use for taking stereoscopic pictures of public ceremonies, street scenes, and seascapes with rolling waves.

This stereoscope is unsigned and undated, but the fact that it has three filters of different colors suggests that it may date from near the end of the nineteenth century. In 1891, Frederick Ives of Philadelphia introduced a color photography process in which three successive images were taken on the same plate through color filters. When the filtered images are superimposed, a full-color picture appears. Ives is known to have made stereoscopic versions of this apparatus, and this model may have been developed from his design.

DUBRONI'S
PHOTOGRAPHIC APPARATUS
PATENTED
60 REGENT STREET 60
Print

Early photography was a messy and complicated business, demanding a working knowledge of chemistry along with infinite patience and a knack with the shutter. The collodion process, introduced by Frederick Scott Archer in 1851, greatly improved the sensitivity of the photographic plate, allowing exposures as short as two seconds for portraiture. But the exposed plates had to be developed immediately—in the field, if necessary—before the wet collodion, a gummy liquid used to hold the photosensitive chemicals on the plate, had time to dry. Landscape photographers, for example, had to carry their chemicals with them and set up tents for processing the plates whenever they wanted to take a picture.

The Dubroni camera outfit, patented by Bourdin of Paris in 1864, attempted to get around some of these problems by processing the plate inside the camera—a foretaste of today's "instant" cameras. His apparatus, shown here with its carrying box and various chemicals, consisted of a roughly spherical chamber with the lens set into the front wall. The back wall was formed by a hinged plate-holder into which a prepared collodion plate could be slid. After exposure, the plate was sensitized and processed within the camera by introducing a series of chemical solutions through a pipette with a rubber bulb mounted on the top. The whole apparatus could be carried in its box and formed an outstandingly compact outfit by the standards of the day. Warmly recommended for amateurs, it achieved some measure of popularity.

However, the camera was not wholly satisfactory, and when Dr. Richard Maddox, an English physician, devised a dry plate alternative based on gelatin and silver bromide in 1871, it quickly replaced collodion. By 1878, improved versions of the gelatin dry plate, which could be stored indefinitely and exposed in a fraction of a second, had relegated cameras like Dubroni's to the museum shelves.

W. & S. JONES
LONDON

The first precision chemical balances were made in the second half of the eighteenth century. The principle of the balance—a beam pivoted in the center, with a pan to hold standard weights hanging from one end and another pan to hold the object to be weighed hanging from the other—is very ancient, dating from the first millennium B.C. But great accuracy only began to be required when gravimetric analysis came into use among the chemists of the eighteenth century, notably the French savant Antoine Lavoisier.

This balance was made in about 1810 by the firm of William and Samuel Jones of Holborn, in London, which operated from 1794 until well into the Victorian era. The Jones brothers built up a flourishing business in scientific instrument manufacture, although few precision balances known to have been made by them still exist. The crossbeam consists of two brass cones joined at their bases, a technique which produces a strong and rigid member. Two spirit levels are mounted on the case of the balance for leveling. In addition, the Jones balance included a set of glass pans, presumably to avoid damage to the brass pans when weighing corrosive materials.

The design of the balance is based on one constructed for the Royal Society in 1789 by Jesse Ramsden, the famous instrument-maker. That balance was used for a study of the specific gravity of water/alcohol mixtures, requested by the Government, which wanted an improved method of levying the excise duty on spirituous liquors. The Jones balance can also be used to measure the specific gravity, or density, of a substance. One of the pans can be replaced by a pan with a glass bob suspended below it, balanced by a brass counterweight in the other pan. Then the glass bob is weighed when immersed in a beaker full of the liquid under investigation. The specific gravity of the liquid could be calculated by the change in weight of the glass bob from a free-hanging state to one in which it is partially supported by the liquid. The denser the liquid, the more strongly it supports the glass bob, and the less the bob weighs in the beaker.

Anscheritz &
Schlaff
London
No 112
Troy
Drams
Dwts.

Justus von Liebig, the nineteenth-century German chemist, called the balance "that incomparable instrument, which gives permanence to every observation, dispels all ambiguity, establishes truth, detects error, or shows us that we are in our true path." He was talking, of course, about the precision chemical balance, whose adoption in the first half of the nineteenth century helped to transform chemistry into an exact science. But there is no doubt that the bank clerk or the money lender who used this commercial instrument felt similarly about his scale.

The self-indicating balance shown opposite was used for weighing gold coins. Made in London in about 1800 by Anscheritz and Schlaff, it stands on a tripod, with the feet individually adjustable by screws for leveling. It has three scales: one measuring in avoirdupois, one in troy weight, and one simply designed to count the number of gold coins in the weighing pan. The mechanical arrangements at the back consist of a simple lever with a pan at one end, a weight at the other, and a pivot in the middle. When coins are placed in the pan, it sinks, turning the three pointers which are attached to the lever so as to register the weight of the pieces. This balance was intended for weighing very small amounts of gold at a time; its maximum was barely over one ounce avoirdupois.

Such a balance is not a precision instrument, but it provided a useful check when gold coins were being exchanged. Coin balances were among the most important items of equipment in treasuries, banks, and in the offices of money changers and businessmen—as long as the value of money depended on the weight of the precious metal it contained. Once coins were no longer made of gold, it mattered less what they weighed, although the modern equivalent of this balance is still used in banks as a quick method of checking the number of coins in a bag of change.

The first instruments which we would recognize as thermometers were made in Florence in the 1650s. Before that there had been instruments which would give a rough idea of temperature changes, using the expansion of air as it was heated to drive a column of water up a tube. But these instruments, known as thermoscopes, were open to the air and were therefore affected by changes in barometric pressure as well as changes in temperature. This made them hard to interpret.

The sealed thermometer, in which a column of liquid expands as it is heated and indicates temperature by the movement of its leading edge up a graduated scale, was invented by Ferdinand II, Grand Duke of Tuscany, in about 1654. The liquid used was alcohol, and the instrument consisted of a glass bulb to which a fine tube was attached. The glassblower Iacobo Mariani made a number of thermometers of this type for the Academia del Cimento in Florence, founded in 1657. The design of the instruments evolved by a process of trial and error. The glassblower would fill up the tube to a certain point with alcohol, heat it to expel the air, and then seal up the end. The use of standard fixed points—the freezing and boiling points of water, for example—was not generally accepted for another century, so in order to make the thermometers more or less consistent one with another, other factors were standardized, such as physical dimensions, quantities of alcohol, and so on. Mariani was a fine craftsman, and his smaller thermometers, which looked very much like those we use today, gave remarkably uniform readings.

The Academia del Cimento was disbanded in 1667, but in 1829 a collection of its glassware was rediscovered; it is now preserved in Florence. This is a replica of the 300° spiral thermometer—designed more for show than for accurate measurement.

It was Evangelista Torricelli, an Italian physicist, who discovered that we live "submerged at the bottom of an ocean of elementary air." He reached this revolutionary conclusion after inventing a device which was able to weigh this invisible column of air—the barometer. He took a four-foot length of glass tubing, filled it with mercury, and then inverted it into a bowl containing more mercury. The mercury level in the tube fell, but it did not all flow out; thirty inches of mercury stayed in the tube. It stayed there, Torricelli reasoned, because of the weight of the air pressing down on the mercury in the bowl. When the weather changed, he noticed, the column of mercury rose or fell according to the humidity. His barometer was the first instrument which could be used to predict changes in the weather.

Torricelli made his first crude barometer in 1643; this more developed example was made more than fifty years later by Daniel Quare of London. The main structure of the barometer is a fluted column of boxwood (of which only the upper part can be seen here), resting on a base of four brass legs. Within the hollow column is a glass tube containing mercury. At the bottom is a mercury reservoir or cistern with a flexible leather base. When the barometer needs to be moved from one place to another, an adjustable screw with a padded head can be tightened so that it presses on the base of the leather cistern, forcing the mercury up completely inside the tube. Thus the heavy liquid is prevented from sloshing around and damaging the glass tubing. This innovation was introduced about 1702 by John Patrick, an instrument-maker who specialized in the manufacture of barometers and was known in London as the "Torricellian Operator."

A constriction in the glass tube near the top is designed to reduce the risk of damage if the mercury should surge up in the tube while being carried. This feature was patented by Quare in 1695 and accounts for the signature "Invented and Made by DAN QVARE LONDON."

31
Rising
FAIR or
FROST
30
VARI-
29
Faling
RAIN, SNOW
or WIND
28
31
DRY
SERENE
30
ABLE
29
RAINY
STORM
28
Invented &
Made By
DAN. QVARE
LONDON

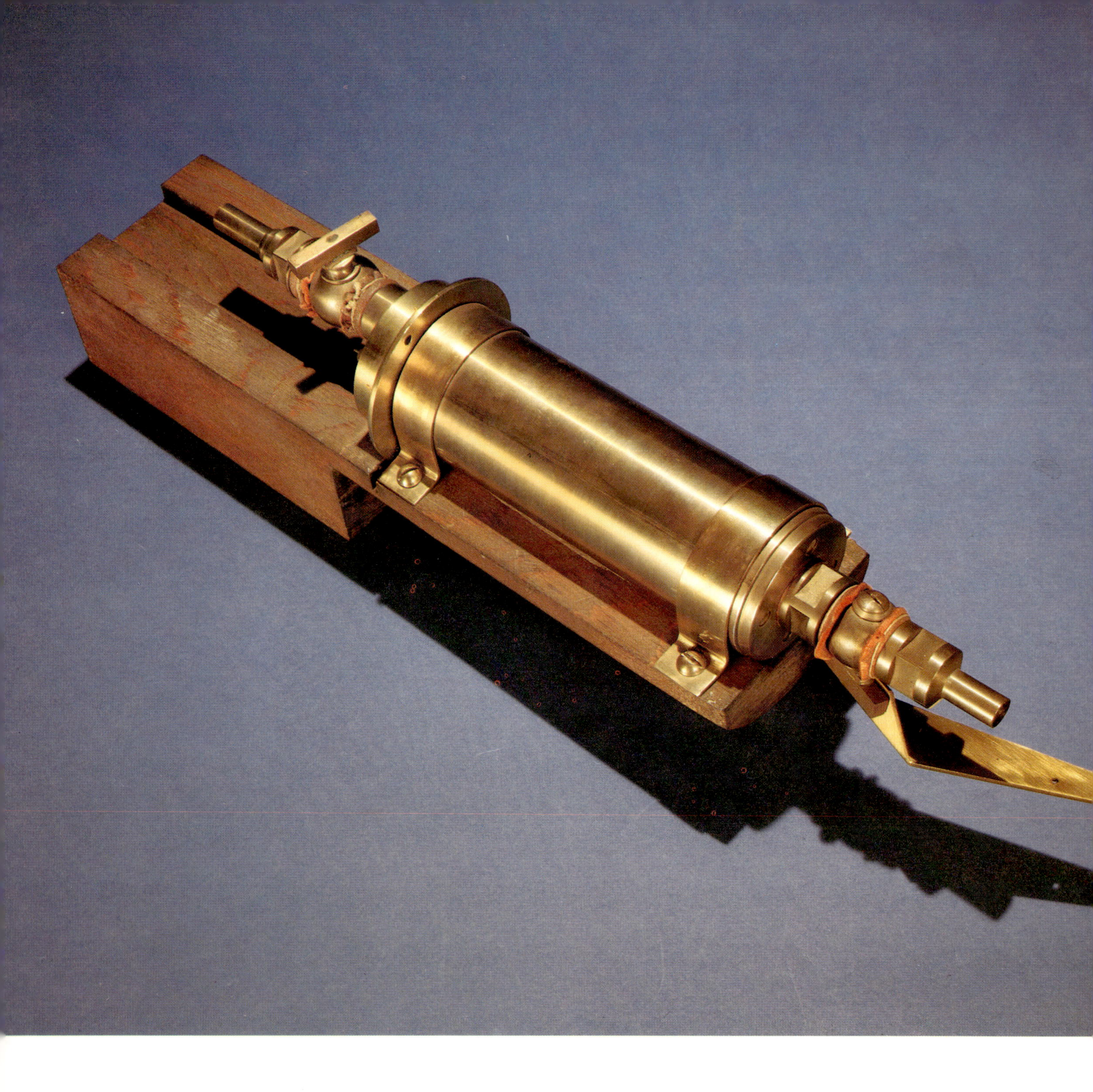

The term eudiometer (from the Greek *eudia*, fair weather) was first used by a minor Italian scientist, Marsilio Landriani, to describe an instrument he used in 1775 to measure the purity of the air. The method had been devised by Priestley in 1772 and involved the combination of a measured volume of air with nitric oxide. In 1777, Alessandro Volta transformed the eudiometer, making it applicable to the general analysis of gases. This eudiometer was used by the eccentric Henry Cavendish (1731–1810), grandson of the second Duke of Devonshire. Although highborn, well-educated, and wealthy, Cavendish lived the life of a recluse, wearing peculiar clothes and devoting himself to scientific investigations. He was terrified of women, and communicated with his female servants by means of notes. "He probably uttered fewer words in his life than any man who lived to fourscore years," remarked Lord Brougham. Nevertheless, he was a remarkable scientist.

Cavendish's eudiometer is simply a vessel in which two or more gases can be made to combine chemically by mixing them and passing an electric spark through the mixture. Cavendish used it to test the purity of the air by mixing samples with nitric oxide and reacting them together. By measuring the volumes of the gases before and after the spark had been passed, he was able to determine the proportion of oxygen in the air. This research led Cavendish to the somewhat surprising solution that "the air of London appeared rather the purest, and sometimes that of Kensington." (Kensington was then a London suburb, but has now been swallowed up by the westward migration of the city.)

More important, Cavendish was the first person to measure the densities of gases and to show that they differed—an important finding in the history of chemistry. He also showed that when hydrogen and oxygen were mixed in the eudiometer and ignited, they were completely converted to water, in an amount precisely equivalent to the combined weight of the two gases. This finding, published in 1784, showed that water consisted solely of hydrogen and oxygen and nothing else, and also revealed the proportions of the two gases in its makeup.

Until the second half of the nineteenth century, virtually all dyestuffs were derived from vegetable or insect sources: red cochineal from the dried bodies of insects, rose madder from an evergreen shrub, saffron from the flowers of the saffron crocus. The first chemical dye was produced serendipitously by William Perkin, a young chemist at the Royal College of Chemistry in London. Perkin had mixed aniline with potassium dichromate in an attempt to make quinine; what he actually made was a black sludge, from which he extracted purple crystals. He found that it would dye silk brilliantly and sent a specimen to Pullars of Perth in Scotland, leading dyers of the day. They were ecstatic: "If your discovery does not make the goods too expensive," they told Perkin, "it is decidedly one of the most valuable that has come out for a very long time." The young scientist went into business manufacturing his dye, which he called mauveine. By the time he was thirty-five, he had made enough money to retire from business and go back to chemical research.

Perkin's discovery set off a whole new industry, and several more new dyes were soon produced; magenta, rosaniline blue, and aniline black. Perkin himself played a major part in the discovery of one of the most important, alizarin, a red dye which is the active ingredient in the age-old dye rose madder. Perkin and a German chemist, Heinrich Caro, discovered how to make alizarin almost at the same time, and rushed to patent the process. Caro beat the Englishman by a day, getting his patent on June 25, 1869. He then sold his patent to the chemical company Badische Anilin-und-Soda Fabrik (more familiarly known today as the chemical giant BASF), and they in turn gave Perkin a license to manufacture alizarin in England.

Perkin's original crystals of alizarin and mauveine are shown opposite, together with some yarn colored with the actual mauveine dye. Although it appears rather lurid by today's standards, mauveine appealed enormously to the Victorians. One of the Pullars wrote to Perkin: "A rage for your colour has set in among that *all-powerful* class of the Community—*the ladies*. If they once take a mania for it and you can supply the demand, your fame and fortune are secure." They were.

COLLEGE
Alizarin
William
LECTURE
ORIGINAL
MAUVEINE
PREPARED BY
SIR WILLIAM PERKIN
IN 1856

BIBLIOGRAPHY

Asimov, Isaac. *Asimov's Biographical Encyclopedia of Science and Technology.* New York: Doubleday, 1972.

Astronomy 1: Globes, Orreries and Other Models. London: Her Majesty's Stationery Office, 1967.

Early Machine Tools. London: Her Majesty's Stationery Office, n.d.

Electric Power Part II: Descriptive Catalogue. London: Science Museum, 1933.

Forbes, Eric G. *The Birth of Navigational Science.* Maritime Monograph no. 10. Greenwich: National Maritime Museum, 1973.

Ford, Brian J. *The Revealing Lens.* London: Harrap, 1973.

Fraser, Grace Lovat. *Textiles by Britain.* London: George Allen and Unwin, 1948.

Gernsheim, Helmut, and Gernsheim, Alison. *A Concise History of Photography.* London: Thames and Hudson, 1971.

Guye, Samuel, and Michel, Henri. *Time and Space.* London: Pall Mall Press, 1971.

Hewson, Commander J. B. *A History of the Practice of Navigation.* Glasgow: Brown, Son and Ferguson, 1951.

King, Ronald. *Michael Faraday of the Royal Institution.* London: Royal Institution, 1973.

Kingsbury, J. E. *The Telephone and Telephone Exchanges: Their Invention and Development.* London: Longmans Green, 1915.

Marland, E. A. *Early Electrical Communication.* London, New York, and Toronto: Abelard-Schuman Ltd, 1964.

Physics for Princes: The George III Collection. London: Her Majesty's Stationery Office, 1968.

Robinson, Stuart. *A History of Dyed Textiles.* London: Studio Vista, 1969.

Rolt, Lionel T. *Tools for the Job.* London: B. T. Batsford Ltd, 1965.

Sewing Machines. London: Her Majesty's Stationery Office, 1970.

Singer, Charles; Holmyrad, E. J.; Hall, A. R.; and Williams, Trevor. *A History of Technology.* Vols. 1–5. Oxford: Oxford University Press, 1958.

Steeds, W. *A History of Machine Tools.* Oxford: Oxford University Press, 1969.

Surveying: Instruments and Methods. London: Her Majesty's Stationery Office, 1968.

Taylor, E. G. R. *Navigation in the Days of Captain Cook.* Maritime Monograph no. 18. Greenwich: National Maritime Museum, 1975.

Temperature Measurement and Control Part II: Descriptive Catalogue. London: Science Museum, 1955.

Time Measurement. London: Science Museum, 1966.

Time Measurement Part II: Descriptive Catalogue. London: Science Museum, 1955.

Ullyett, Kenneth. *In Quest of Clocks.* Feltham, Middlesex: Spring Books, 1968.